The Horse Who Came to Dinner
The First Criminal Case of Food Fraud

The Horse Who Came to Dinner
The First Criminal Case of Food Fraud

Glenn Taylor
Hampshire Scientific Service, UK
Email: grjtaylor@icloud.com

Print ISBN: 978-1-78801-137-2
EPUB ISBN: 978-1-78801-712-1

A catalogue record for this book is available from the British Library

© Glenn Taylor 2019

All rights reserved

Apart from fair dealing for the purposes of research for non-commercial purposes or for private study, criticism or review, as permitted under the Copyright, Designs and Patents Act 1988 and the Copyright and Related Rights Regulations 2003, this publication may not be reproduced, stored or transmitted, in any form or by any means, without the prior permission in writing of The Royal Society of Chemistry or the copyright owner, or in the case of reproduction in accordance with the terms of licences issued by the Copyright Licensing Agency in the UK, or in accordance with the terms of the licences issued by the appropriate Reproduction Rights Organization outside the UK. Enquiries concerning reproduction outside the terms stated here should be sent to The Royal Society of Chemistry at the address printed on this page.

Whilst this material has been produced with all due care, The Royal Society of Chemistry cannot be held responsible or liable for its accuracy and completeness, nor for any consequences arising from any errors or the use of the information contained in this publication. The publication of advertisements does not constitute any endorsement by The Royal Society of Chemistry or Authors of any products advertised. The views and opinions advanced by contributors do not necessarily reflect those of The Royal Society of Chemistry which shall not be liable for any resulting loss or damage arising as a result of reliance upon this material.

The Royal Society of Chemistry is a charity, registered in England and Wales, Number 207890, and a company incorporated in England by Royal Charter (Registered No. RC000524), registered office: Burlington House, Piccadilly, London W1J 0BA, UK, Telephone: +44 (0) 20 7437 8656.

Visit our website at www.rsc.org/books

Printed in the United Kingdom by CPI Group (UK) Ltd, Croydon, CR0 4YY, UK

Foreword

The Horse Who Came to Dinner is a unique book for several reasons. It is a mature reflection on the horsemeat scandal, linking what happened then to previous food frauds and bringing us up to date with the aftermath and current problems. In Glenn Taylor, we have an author with high level enforcement experience, a 'helicopter-view' of policy and a flowing narrative style. Thus, this book gives us fresh insights into global threats, the legislative context, emerging technologies and latest thinking on food fraud prevention. There are also informed perspectives on food crime and the sanctions provided by the courts with relevant examples of significant cases. Glenn does not shy away from vexed questions on the implications for food fraud of UK exit from the European Union (EU) and new models of food regulation.

The harm that food fraud inflicts is well known; consumers can be duped into spending money unnecessarily, which increasing numbers can ill afford, trust is lost and business reputations are damaged. Food safety may also be compromised; for example, the deaths that have arisen from counterfeit alcohol products and fraudulent activity in the allergen supply chain. 'Horsemeat – 2013' was a watershed. The subsequent Review by Professor Elliott, in which I was privileged to play a part, described how 'cutting corners' elides through food fraud into '*food crime*', an organised activity by groups that knowingly set out to deceive, and/or injure, those purchasing food.

This book reviews historical food fraud and questions if people realise that even modern cheap food can be the subject of fraud. The successes and failures of the Elliott Review are discussed, along with risk assessments of topical issues such as sugar and healthy eating. Significant frauds – melamine in baby milk, the Spanish olive oil scandal – are analysed alongside fipronil in eggs and other contemporary food frauds. Court cases such as *R v Farmbox*, *R v Flexi Foods* and *R v Zaman* are discussed in detail. The need for adequate sanctions and new sentencing guidelines is reviewed.

The difficulties of regulating the global marketplace are exemplified with a discussion of fraud in the food supplements marketplace, bringing in the 'thinking like a criminal' mindset and the use of intelligence to foil fraud attacks and bolster defence strategies. There are lessons to be learned from other sectors, including the 2008 banking crisis and roles played by investors and insurance. The gatekeepers of food authenticity, self-enforcement initiatives and a critical analysis of the EU enforcement system are welcome features of the book with insights into working with major transgressor nations to improve their performance.

The need for technical innovation, industry and enforcement joint working and analytics and the power of big data form important sections of the book and, as is to be expected of the author, insights into what the future may hold.

It is an honour to be asked to contribute a foreword to this significant addition to the literature on food fraud, and I congratulate Glenn for a splendid book which deserves to be widely read.

Michael Walker MChemA, FIFST, FRSC
Expert witness, Referee Analyst and Head of the
Office of the Government Chemist

Preface

In 2013, a major crime was detected. Thousands of consumers, retailers and food businesses were ripped off by thieves who sold horsemeat in place of beef, which resulted in one major food brand ceasing to exist and millions wiped off the stock-market values of major food businesses, a major recall and a major loss of consumer confidence. The meat was traded across the European Union and the network of food companies and brokers involved made it much more difficult to trace who the perpetrators were. It has taken nearly five years to bring the case to court. A horse had literally come to dinner throughout the UK, and a thief, or rather a group of thieves, had stolen from their customers by defrauding them. The public and the UK Government are smarting and looking for strong justice.

    The title of this book, 'The Horse Who Came to Dinner', seemed entirely appropriate. Was this simply a grammatically incorrect play on words or an accurate reference within the archive of my mind? I remembered the 1973 comedy film which was loosely based on a novel by Terrence Lore Smith, 'The Thief Who Came to Dinner'. The film featured Ryan O'Neal as Webster McGee, an honest computer programmer by day and a jewel thief by night. He gathered 'intelligence' by infiltrating the wealthy Texans, gaining their trust and forming a network of

---

The Horse Who Came to Dinner: The First Criminal Case of Food Fraud
By Glenn Taylor
© Glenn Taylor 2019
Published by the Royal Society of Chemistry, www.rsc.org

other insiders, which he hoped would help cover his tracks, thus avoiding prosecution and maximising gain. The dilemma in the film was whether his new-found socialite friends would reveal his secret to the authorities, and would he pay for the crimes he committed. There are too many similarities between the two stories for me to think that this was purely coincidental. There has to be subconscious filtering occurring within my mind; some would say not an everyday event in my case!

No industry, and particularly one like the food industry that relies on consumers trusting the brand, wants to admit to wholesale fraud.

In the horsemeat scandal, the food brokers were the Webster McGees, the criminals who used their inside knowledge and expertise within the food chain to defraud their wealthy clients (large food businesses). Through a network of suppliers across Europe, they hoped to cover their tracks and continue as a trusted supplier within their target group. Fortunately, the only injury was financial; yet, this was no comedy as it has resulted in wholesale change in the way UK enforcement will investigate and, when necessary, prosecute food crimes as a criminal offence. If defendants are found guilty of food fraud in future then Horsegate is the ground-breaking example to be referred to, and I expect the courts will give stiffer sentences to deter other potential offenders. Certainly, by treating this crime as fraud and not adulteration, the courts have greater powers as a result of increased penalties and new sentencing guidelines introduced following these ground-breaking cases. This has changed the way enforcement 'does business' once and for all. Food fraud is now being prosecuted as a crime and not adulteration, and this is much more likely to lead to imprisonment than a simple fine.

You would be forgiven for thinking food fraud is a new concept but records list concerns during the Roman and Greek eras and regularly thereafter. During the mid-1800s, food crime was rife in London and probably throughout Europe. In fact, it was argued that it was almost impossible to buy food which wasn't contaminated. Sawdust was an ingredient of choice, added to bread alongside chalk to increase the weight of a loaf and added to coffee which was already 'cut' with chicory. Since 1860, this practice was referred to as simple adulteration and, as a result, carried less threat to the perpetrators. Consequently, the regulators took to naming and shaming traders as this harmed their business. Those caught selling substandard foods were shamed into improving their act with a strong possibility of a loss of public confidence and therefore customers. It wasn't necessary to prove a fraudulent act or intent, just name and shame.

In response to Horsegate, the government commissioned a report by Professor Chris Elliott from Queen's University Belfast. This led directly to the establishment of a food crime unit in the Food Standards Agency (FSA) – the government body responsible for ensuring the safety of our food. In fact, food crime and food fraud became the focus rather than adulteration. The government has shown an ambition to clamp down on food fraud and has introduced new sentencing guidelines for judges to improve consistency of approach when dealing with food fraud convictions. Where there is sufficient evidence to prove a fraud offence, the offender will be charged with an offence that can be heard in the Crown Court and, if convicted, the judge can impose a greater penalty. Magistrates will still hear food adulteration cases but the range of sentences available to them is much less punitive.

Do we expect fraud if a product is cheap? If you buy a Rolex watch that normally retails for around £4000 from a trader at a ridiculously low price, for example £10, do you expect it to be genuine? No, you are complicit and 'accepting of the fraud'. Often these watches are traded as fakes, but not always. Take the example of sellers on beaches in holiday resorts who sell 'copy' watches without any description whatsoever. Whilst, amongst other things, this is an infringement of copyright, the buyer seems happy to purchase the goods in the knowledge that something isn't right. You could argue that they are complicit and perhaps that situation changes with the price paid. So, is it fair to say that, the higher the cost, the more likely the consumer expects the description to be accurate? If the shop resembles an expensive jeweller's shop and charges £2000, you don't expect a fake and would expect the police to take action if it transpired you had been sold a fake.

Food is different; irrespective of the price, the consumer expects the description to be accurate and true. For over 150 years, food legislation has stated that food must be of the nature, substance and quality so demanded by the consumer. Therefore, no matter how cheap the food, the consumer expects the product to be genuine, and when major food businesses are involved and food is found to be fake, then 'heads should roll'. Now politicians have 'upped their game' and new guidance for judges has led to potentially much longer sentences.

Fraud isn't just a case of duplicity on the side of the criminal, it is aided and abetted by buying systems which enable it to flourish, and thus, purchasers are accepting of fraud by failing to undertake suitable and sufficient monitoring. Fraudsters gain the trust of their target in order to defraud them, often having inside knowledge of systems which enable them to go undetected. The one up, one down traceability monitoring system aids this when complex supply systems are used. There is evidence that fraudsters try on a small scale first and then, once successful, scale up the operation. Fraud is often a high volume, low value crime which is notoriously difficult to detect. Consequently, it goes unnoticed – that's the whole point of fraud, it goes unnoticed for a long time. In markets where supply chains alter frequently, fraudsters operate using verbal contracts which results in ambiguous standards and specifications, and this method of 'doing business' supports fraud.[1] Therefore, food businesses need ever-changing defence strategies to protect them against fraud.

The Department for Environment, Food and Rural Affairs (DEFRA) estimated in 2015 that the national spend on food was £201 billion.[2] A report published at the time suggests that fraud levels are in the vicinity of 6% of turnover.[3] If that is the case for food, then food fraud amounts to a staggering £11 billion per annum, and the vast majority is not reported and perhaps remains completely undetected. At the height of the horsemeat crisis, when the media were hunting for scalps, the boss of Iceland, Malcolm Walker, made some interesting comments:[4]

> ...the horse meat scandal was a 'storm in a tea cup', and insisted that supermarkets should not be blamed for it.
>
> Most supermarkets sold three types of products, he said – premium, standard and economy – and he would not eat the economy product.

> *'Iceland has never sold economy products – we do not sell cheap food.'* he said
>
> *'We have one brand of food, one level of food... we know where all our food comes from, we follow the supply chain right the way through and it's very short.'*
>
> Mr Walker, whose chain has removed beef products which tested positive for traces of horse DNA, denied that horsemeat was endemic in the food chain.
>
> He insisted Britain's supermarkets shouldn't be blamed for contamination and claimed they had *'a fantastic reputation for safety' of products'*.
>
> Instead, he blamed local councils for the scandal, saying they had compromised food quality by trying to drive down the prices of catering contracts for schools and hospitals.
>
> Mr Walker said, *'Supermarkets carry out an enormous range of testing procedures on every product that bears their name.*
>
> *Okay, you can say we haven't been testing for horse. Well, why would we? We don't test for hedgehog either'.*

History shows that previous food scandals are quickly forgotten by the public, and the suggestion that buyers get what they deserve because they demand cheap food has been raised before, but the fake watch scenario does not hold for food. During the horsemeat crisis, I asked fellow regulators across the UK if they had tested for horsemeat during the previous year. Only the Welsh public analysts confirmed they had tested in nine cases and the results were negative. Scottish and English public analysts did carry out some tests. For example, did beef contain sheep, but not horse or hedgehog. The problem with the DNA test used by most public analysts at the time was that the analyst must look for a particular species to determine its presence or absence. As a result, beef containing 70% horsemeat would test positive for beef if it was assessed for the presence of beef.

Everyone started testing for horse immediately after the event – shutting the stable door after the horse had bolted! It seems the only people with 'intelligence' were the fraudsters. The regulators and those undertaking monitoring, including their advisers, were unaware of the threat of horsemeat substitution. Therein lay the dilemma; a fraudster knowing this, or merely having tried to defraud on a low scale and 'got away with it', could pass-off horse in beef for his advantage and continue to do so until those checking 'woke up to this latest crime'. Hence the need for ever-changing defence strategies and enforcement waking up to changing threats.

Malcolm Walker criticised local authorities for encouraging fraud by having an unrelenting focus on the cheapest price and nothing else. The authority I worked for was, to my knowledge, one of perhaps only two in the UK undertaking public analyst checks on food supplied to our schools. We had tested meat species to identify the type of meat supplied and rejected previous consignments following the presence of other species, but we hadn't tested for horse. In addition, we undertook many other tests and audits and regularly found things which didn't meet our exacting specifications. We were well known amongst our suppliers for stringent testing of food, and I suspect that helped maintain high standards as our suppliers didn't know what we tested for, just that we regularly monitored the food they supplied. We weren't supplied with horsemeat, but schools in neighbouring authorities were. It is interesting to speculate that, just like us, large food businesses with a reputation for the most stringent checks were not found to be supplied with horsemeat. Was that luck, or did the fraudster know and seek to avoid detection by going elsewhere?

REFERENCES

1. Portsmouth University professor leads bid to restore trust after horsemeat scandal, http://www.portsmouth.co.uk/news/education/portsmouth-university-professor-leads-bid-to-restore-trust-after-horsemeat-scandal-1-6283310, [accessed August 2018].
2. Food Statistics Pocketbook 2015, in Year Update, https://www.gov.uk/government/uploads/system/uploads/attachment_data/file/526395/foodpocketbook-2015update-26may16.pdf, [accessed August 2018].
3. The Financial Cost of Fraud 2015, What the latest data from around the world shows, http://www.pkf.com/media/31640/PKF-The-financial-cost-of-fraud-2015.pdf, [accessed August 2018].

4. 'I don't eat value food because it doesn't contain much meat': Boss of budget supermarket Iceland in shock claim as Waitrose chief warns of cheap food risks, http://www.dailymail.co.uk/news/article-2280005/Horsemeat-Boss-budget-supermarket-Iceland-shock-claim-Waitrose-chief-warns-cheap-food-risks.html, [accessed August 2018].

About the Author

The author's working life has been dedicated to detecting and stopping crime, researching and sharing defence strategies against food crime in particular. With over forty years in the field, he is an internationally recognised expert who, over the past ten years, has presented at major food enforcement events throughout the European Union (EU). For ten years, he directed a leading EU official control laboratory (public analyst) providing national and international scientific, advisory and analytical services ranging from food protection to the provision of scientific support to Customs, Coroners, Environmental Health, Police and Trading Standards. He has provided expert witness evidence in English courts at all levels, giving him a unique insight into the court system and how to learn lessons to create enforcement with real impact. His national contribution has involved being a board member for Chartered Institute of Environmental Health's Food Board (CIEH), sitting on food intelligence and emerging risk boards within the UK Food Standards Agency, and his research interest is the efficacy of food enforcement and, in particular, the development of effective intelligence. He has produced a number of peer-reviewed papers with fellow scientists from many establishments and collaborated with Government Chemist, Manchester, Kingston,

---

The Horse Who Came to Dinner: The First Criminal Case of Food Fraud
By Glenn Taylor
© Glenn Taylor 2019
Published by the Royal Society of Chemistry, www.rsc.org

Southampton and Portsmouth Universities, and he has a book on forensic enforcement published by the Royal Society of Chemistry. This book reflects his inside knowledge and experience. He was there, in the thick of it, when Horsegate happened and that gave him a unique insight.

This book is dedicated to my three grandsons.

Joshua, Reuben and Robin.

My generation made a mess of finance which resulted in one of the longest and hardest periods of austerity ever recorded. We were not all responsible, yet in some way, we played our part. As we try to sort out the mess, we reduce investment in too many areas.

We now have an opportunity to leave something more positive for your generation: safe food in its widest sense. The world is changing faster than it has ever done before. The Internet will impact on all you do, and we need to be smart and ensure enforcement supports the law-abiding of your generation by giving them the sound, science-based information upon which to make informed decisions.

I hope we leave the right legacy and one of which we can be proud.

With heartfelt love

Contents

Chapter 1
The Luck of the Irish — 1

1.1 A New Dawn — 1
1.2 How to Commit Fraud in the Meat Trade — 3
1.3 The Investigation — 5
1.4 There is Nothing Like Being Caught on the Hop — 7
1.5 The Ground-breaking Prosecution Cases — 9
    1.5.1 First UK Prosecutions: Boddy and Moss – Southwark Crown Court March 2015 — 9
    1.5.2 The Second Case: Raw-Rees and Patterson – Southwark Crown Court May 2015 — 12
    1.5.3 The Main Horsemeat Scandal Case — 14
1.6 The Luck of the Irish — 16
1.7 Is There Another Case Yet to be Brought to an EU Court in the Future? — 16
References — 18

Chapter 2
The Grecian Temple of Food Enforcement — 21

2.1 Eight Pillars to Support the Food Enforcement Temple — 22
2.2 So What Were the Eight Pillars and What Progress Has Been Made Since His Report? — 24

The Horse Who Came to Dinner: The First Criminal Case of Food Fraud
By Glenn Taylor
© Glenn Taylor 2019
Published by the Royal Society of Chemistry, www.rsc.org

	2.2.1	Pillar One: Consumer First	24
	2.2.2	Pillar Two: Zero Tolerance	28
	2.2.3	Pillar Three: Intelligence Gathering	29
	2.2.4	Pillar Four: Laboratory Services	31
	2.2.5	Pillar Five: Audit	37
	2.2.6	Pillar Six: Government Support	39
	2.2.7	Pillar Seven: Leadership	40
	2.2.8	Pillar Eight: Crisis Management	41
2.3		Was it a Success?	42
References			43

Chapter 3
Is Food Fraud a New Idea?   44

3.1	Did the Public Knowingly Accept Adulterated Food?	50
3.2	Synthetic Food and Adulteration	51
3.3	Has Adulteration Changed Since the 19th Century?	52
3.4	Fraud or Adulteration?	53
	3.4.1 Definitions	54
	3.4.2 The Non-compliant Products	56
References		58

Chapter 4
World-wide Food Frauds   60

4.1	Beef Pies Supplied for UK County Catering Establishments	60
4.2	Fipronil in Dutch Eggs	62
4.3	German Asparagus	66
4.4	Chinese Milk Scandal	68
	4.4.1 Why Add Melamine to Food?	68
4.5	Olive Oil	71
	4.5.1 The Spanish Olive Oil Scandal	72
4.6	French Wine Fraud	75
	4.6.1 Breaking News	77
4.7	Fish Fraud in the United States and World-wide	78
4.8	Chilli Powder From India	82
References		82

Chapter 5
Off With Their Heads   86

5.1	The New Guidelines	87

5.2	Timetable of Events	87
5.3	World-wide Harmony of Food Law	89
5.4	The FDA Food Safety Modernization Act (FSMA) United States	91
5.5	Amazon	92
5.6	Adulterated Food	93
5.7	The Elusive Due Diligence Defence	95
5.8	A New Dawn in the UK	98
5.9	Clarity For Those Involved in Prosecuting	98
5.10	In Summary	99
	5.10.1 ASDA	100
	5.10.2 Sideras, Nielsen and Ostler-Beech	100
	5.10.3 Boddy and Moss	100
	5.10.4 Zaman	100
References		101

Chapter 6
Mission Impossible; Regulating a Global Marketplace   103

6.1	Food Supplements: 'The Gym Bunny's' Little Helper!	105
	6.1.1 Definitions for Medicines and Supplements	105
	6.1.2 Is UK Enforcement Doing Enough?	106
6.2	Sildenafil Citrat	111
6.3	So What Does This Have to do With Food Fraud?	112
References		117

Chapter 7
Thinking Like a Food Fraudster   120

7.1	Overview	120
7.2	Legislation	120
7.3	Attack	125
	7.3.1 The People Involved	126
	7.3.2 The Foods Involved	127
7.4	Defence Strategies	132
	7.4.1 The Importance of Communication: A 'Burglar Alarm on Britain'	132
	7.4.2 A Vital Role for Government (Member States)	133
	7.4.3 A Key Role for Industry	134
	7.4.4 The Weakest Link	136
References		137

Chapter 8
The Bank Job 139

8.1 Will the Major Food Businesses and Regulators Learn From the Banking Crisis? 139
References 143

Chapter 9
Someone to Watch Over Me 145

9.1 TSB: The Fraudster's Eyes Are Watching You 150
9.2 Who Watches EU Food Safety? 151
9.3 A Comparison of the Top Three Gatekeepers of EU Food Safety from 2003–2017 155
    9.3.1 Italy 155
    9.3.2 UK 155
    9.3.3 Germany 155
9.4 Summary 158
References 160

Chapter 10
Look What They've Done to My Song 162

10.1 So Much Change 162
10.2 Complex Food Legislation and a Lack of Harmony Across the Globe 163
    10.2.1 Lack of Harmony in Legislation 165
10.3 Inconsistent Response from Enforcement 166
10.4 Risk-based Enforcement 168
10.5 World-wide Web of Food Supply 170
    10.5.1 Increasingly Cosmopolitan Diets and a More Complex Food Supply Chain 170
    10.5.2 More Complex Food Supply Chain 170
    10.5.3 Is There an Insatiable Appetite for Cheaper Food? 171
10.6 Austerity, Increasing Demand and Reducing Spend on Enforcement 171
10.7 The Role of the Media in Reporting Non-compliance and the Reaction to These Reports by the Public and Politicians 173
10.8 Relationships Between the Key Stakeholders 175
    10.8.1 Consumers 175

Contents   xxi

	10.8.2 Regulators and Enforcers	175
	10.8.3 Food Businesses	176
	10.8.4 Politicians	176
10.9	The Potential for Confusion, Especially Regarding Who Is Responsible for Food Safety Failures	177
	10.9.1 Who Is Responsible for Food Safety?	178
10.10	Post-Brexit Enforcement	178
10.11	One Voice	179
References		180

Chapter 11
New Kids on the Block   183

11.1	The Need to Innovate	183
11.2	Exciting New Techniques	184
11.3	How Do You Use Intelligence to Defend Against Food Fraud?	189
11.4	Analytics Early Detection of Emerging Data	191
11.5	WWW Data Sharing Research	194
	11.5.1 Use of Emerging Signals to Predict a Trend	195
	11.5.2 Provenance	195
11.6	Challenges to Using Intelligence	197
11.7	Superheroes of Food Enforcement	198
	11.7.1 The Procrustean Bed of Food Enforcement	198
	11.7.2 Is it Time for a Rethink?	200
11.8	How Do We Move Forward?	202
References		203

Chapter 12
It Ain't What You Do It's the Way That You Do It   205

12.1	How Could Enforcement Improve?	207
References		208

Chapter 13
Who Have We Invited to Dinner Next?   210

13.1	So Who is Coming to Defraud You Next?	210
13.2	Is the Pursuit of Cheaper Food to Blame?	211
13.3	Should We the Customer Buy Directly From the Other Side of the World?	212

13.4	Follow the Money	212
13.5	So What Products Are Next?	213
13.6	Is There Any Guidance to Stop This?	214
13.7	Back to the Beginning of My Career	215
13.8	What Will Happen Post Brexit?	216
References		217

Subject Index   218

CHAPTER 1

The Luck of the Irish

1.1 A NEW DAWN

The world changed with the discovery of horse passed off as beef in 2013. Well, at least the enforcement world in the UK changed; it would never be the same again, and this would impact across the European Union (EU) world of enforcement too. Why? Horse has been passed off as other meats before, and there is a history of other species, for example lamb and goat, being sold as other more expensive meats too. Generally, the public is apathetic about these food scandals; they come and they go. The media kick up a little storm, and then the food scandal fades from the memory. Food seems to be treated differently; food fraud wasn't called fraud, it was called by many other names but not fraud. So why was this case different? Horsemeat was passed off as beef in foods sold in Sweden, France and the UK. The whole of the food industry was caught off-guard and this caused a reaction amongst the public and particularly politicians, who responded to media and public opinion. How dare someone pass off horse as beef! This scandal reached the very top, even the Prime Minister was involved. The perpetrators will feel the full force of the law. Was this the start of a new era in food enforcement?

Substitution of beef with horse became a major crime; at the time, it was not on the horizon of virtually anyone involved in food either as a business, retailer, manufacturer, supplier or the enforcement agencies. It should have been, and the Irish were alive to the possibility. The fraudsters discovered an opportunity to use a plentiful supply of cheaper horsemeat and substitute it for the more expensive beef and make a financial killing; and they took full advantage – the perfect fraud? The disguise was simple; mince the meat and sell it as minced beef. Thousands of consumers, retailers and food businesses were ripped off by these thieves. They used a web of deception, buying horsemeat and beef from separate suppliers, mixing the two, then passing the combined horsebeef mince off to the 'trusted suppliers' of some of the largest retailers and food businesses to sell on the fraudulent produce to their clients who were all naive to the issue. They successfully duped everyone, including the consumer.

The perfect fraud was helped by the fact that those checking the mince would need to carry out testing for a specific species, in this case horse, as there was no simple test to check for all species. Testers needed intelligence or a hunch to know that they should test for horse. Early in 2013, I asked public analyst colleagues throughout the UK if any were looking for horse in the previous year. With two exceptions, all replied no although a few looked for substitution of pork, lamb and beef species. One of the Welsh public analysts did test for horse, and their nine samples gave negative results. The Irish authorities didn't follow the enforcement crowd as their routine investigations looked for horse (using the other public analyst), and they found horse where beef was declared. Did the fraudsters know that the testing regimes helped their fraudulent activity go undetected? Did they have inside knowledge or were they just lucky and had access to an opportunity; namely, to buy and mix horsemeat and beef?

Food fraud was a whole new ball game as, previously, adulteration of food was the preferred charge for regulators. This simply needed analytical proof that the food was not compliant with the regulations, *i.e.* checking meat species and confirming that pork was not present when beef mince had been declared. Food enforcement officers were comfortable with this approach as it involved straightforward analysis and, if necessary, a routine prosecution through the Magistrates Court. It was a reliable system which had served them well since 1860. The old reliable system may have worked well, but it didn't focus on fraud or finding the brains behind a fraud – the criminals; it simply focussed on proving that food was not of the required standard at the point of sale. This resulted in fines, often relatively small. Did this make the fraud worth the risk; after all, the gains could be far more significant than any fines? The risk of getting caught was hardly an issue if the fines were small and the enforcers were looking elsewhere. Something needed to change.

Yet, it has to be said that to prosecute the offence of fraud is far from straightforward. It is necessary to demonstrate that the defendant's actions involved five separate elements:

1. false statement of a material fact;
2. knowledge on the part of the defendant that the statement is untrue;
3. intent on the part of the defendant to deceive the alleged victim;
4. justifiable reliance by the alleged victim on the statement; and
5. injury to the alleged victim as a result.[1]

Trying to gather evidence to prove intent was a complete departure from the norm for food enforcement officers. Police colleagues were needed to maximise the chances of a successful prosecution. Politicians wanted action and demanded that the key culprits were found and prosecuted. A new dawn in public protection arrived, and would food enforcement ever be the same again?

1.2 HOW TO COMMIT FRAUD IN THE MEAT TRADE

Meat is a commodity which can be traded like any other and is classified as follows:

- primal cuts: these are the best cut from an animal – for example, loin or rib and sirloin – and may be sold whole or cut further. These are harder to forge as experienced butchers may be able to identify different species by sight;
- trim: the meat removed from a carcass after the primal meat is removed. It is often minced and used for manufacturing

burgers, for example. Trim is sold using a numerical classification such as 80/20, lean meat and fat contents, respectively; and,
- mechanically separated or recovered meat: as its name implies, this is the lowest quality of meat removed from the carcass. It is forced through sieves using high pressure to separate the bone and produce a 'meat slime or paste'.

The commodity can be traded between many brokers without leaving the original slaughterhouse and is eventually sold to a manufacturer for processing. This is because many brokers do not have the facilities to store meat in large quantities. Many in the food industry believe that, the longer the line of brokers, the more the risk of fraudulent activity as this increases the potential for a web of deceit. The current defence against fraud of 'one up, one down traceability', espoused by some food businesses and recommended by regulators, may aid this. As the name implies, the principle is that each member of the chain (sometimes referred to as a web to demonstrate the complexity and world-wide nature of the process) is responsible for ensuring the traceability of the product from the adjacent members within the chain. My employers and some of the largest retailers moved away from this approach a few years before Horsegate and demanded very short supply chains so that the manufacturer personally knew the origin of the food. This builds closer partnerships and dramatically reduces the likelihood of fraudulent activity.

In the case of a fraud involving beef and horse, the two meats must be combined, which more than likely means sourcing meat from two slaughterhouses. This is to keep the number of knowledgeable people to a minimum and process the meat within the chain utilising an unsuspecting 'trusted broker' with an excellent reputation to sell to the food manufacturer. A schematic for such an operation might be as follows:

Slaughterhouse A ⟶ Broker 1 ↘
            Broker 3 (processing plant) ⟶ Broker 4 ⟶ Food manufacturer
Slaughterhouse B ⟶ Broker 2 ↗

Slaughterhouse A might be the supplier of beef and Slaughterhouse B the horse.

The slaughterhouses, other brokers and manufacturer would be blissfully ignorant of the fraud and the fraudster would seek to keep it that way, trying to ensure that as few people as possible were alerted.

At every stage, documentation must support the consignment and be available for inspection when required, so to disguise a fraud, the documentation must be altered to 'cover the tracks of the fraudster'. This type of forged paperwork starts to build the required elements for the full offence of fraud.

Horsemeat can be legally traded throughout the EU; in some member states, it is prized for human consumption (France, Hungary, Italy and Belgium). In others, such as the UK, it is not a meal choice, possibly because the UK public see a horse as a pet and find it abhorrent that it could be served up on a dinner plate. There are regulations concerning abattoirs and the processing of horse and beef to ensure that the two cannot be 'inadvertently' mixed, and the precise nature of the meat must be as declared on the label provided by the slaughterhouse.

1.3 THE INVESTIGATION

During 2012, two investigations being carried out by enforcement officers of the Food Standards Agency Ireland (FSAI) were a survey at wholesalers of beef products and samples of foods taken at food retailers (market surveillance). The two initiatives run by the Irish authorities were not connected.

The investigation which looked at wholesalers eventually led to a sizeable fraud prosecution and arose from a random sampling visit to Freeza Meats Ltd in September 2012 by environmental health officers from Newry and Mourne District Council.[2] This time, pallets of frozen meat trim were found which were being stored on behalf of McAdams Foods. The labels suggested the meat had originated from Poland and it was being unloaded at McAdams. In fact, this was a lie as the meat had been unloaded elsewhere and the documentation was falsified to hide this information. Further checks on the consignment in October 2012 and early 2013 revealed an identification chip from an Irish pony and a polish horse called Victor.[3] The pallets were tested and proved positive for horse and beef DNA in various mixes.

The retail survey was linked to a series of campaigns which had run across a range of foodstuffs over the previous seven years, assessing what was available for sale on the high street. This sort of investigation has been carried out by many enforcement officers across the UK, although the numbers are reducing dramatically as budget cuts bite. It is interesting to note that the Irish should be praised for not slavishly following a system of risk-based monitoring which doesn't incorporate street surveillance, *i.e.* 'just checking' what is available for sale on-the-market through retailers. In recent years, risk-based monitoring has become in-vogue as it allegedly reduces the burden on food businesses and the workload for short-staffed government enforcement agencies that might be strapped for cash. Had this been followed slavishly, horsemeat fraud would never have been detected.

In 2012, the retail focus was on meat products such as foods containing beef, for example lasagne, salami and beefburgers, and checking them for authenticity using sophisticated DNA techniques to test for pork, horse and beef.[4] The first positive result from this survey looking for horse DNA was revealed on 10 December 2012.[4] Consequently, more samples were taken for DNA profiling on 18 and 21 December 2012.[4] The samples were sent to IdentiGen (Ireland) and Eurofins (Germany) with an additional request to quantify the amount of horsemeat found. Results were received on the 11 January 2013, and these confirmed that nine out of the ten burger samples submitted tested for low levels of horse and that the tenth contained around 29% horse. The product concerned was beefburgers manufactured by Silvercrest for Tesco.[5]

On 14 January 2013, the FSAI informed the Department of Health and the Department of Agriculture, Food and the Marine in Ireland of the final results of the retail survey. On the same day, it also informed the Food Standards Agency (FSA) in the UK.[4]

The extensive testing regime yielded the following results:

'*Of the 31 beef meal products (such as cottage pie, beef curry pie or lasagne), all were positive for bovine DNA, 21 (68%) were positive for porcine DNA and none were found to contain equine DNA. Only two of these beef meal products declared on the label that they contained pork, which was found at very low levels and therefore we considered its presence may be unintentional and due to cross-over from processing of different animal species in the same plant.*

Of the 27 burger products analysed, all were positive for bovine DNA, 23 (85%) were positive for porcine DNA and 10 (37%) were positive for equine DNA. Most of the burgers positive for porcine DNA were not labelled as containing pork which was found at very low levels and again we considered its presence may be unintentional and due to cross-over during the processing of different animal species in the same plant. The 27 burgers which were tested in this study came from nine different manufacturers, six in Ireland and three in the UK. The products which tested positive for equine DNA came from three plants, two in Ireland and one in the UK.[4]

The five retailers concerned were Tesco, Dunnes Stores, Aldi, Lidl and Iceland, and the next day, 15 January 2013, the FSAI advised them all of their findings, resulting in all of these firms withdrawing the offending products.[4] The media and newspapers of 16 January 2013 in Ireland and the UK led with the story, focussing on the one burger which tested positive for 29% equine DNA.[6] Ten million burgers were instantly removed from sale. Several other food businesses removed frozen meat products as a precautionary measure.[7] Ireland's Department of Agriculture published findings in March 2013 which *'concluded that there is no evidence that Silvercrest knowingly purchased horsemeat'*.[8]

1.4 THERE IS NOTHING LIKE BEING CAUGHT ON THE HOP

The results of this investigation were released and they caused a media furore and grabbed public attention more than any previous food scandal. Concern was expressed at very senior levels and Horsegate headlines could not be missed. The Prime Minister was rumoured to be taking an active interest.

On Sunday 10 February 2013, whilst serving pre-lunch drinks for friends at our home, I was telephoned by a senior manager at the FSA who informed me that they had to update the Prime Minister's office at 3 o'clock that afternoon on the progress made in the fight to detect horsemeat fraud. He was apologetic for disturbing me on a Sunday but needed advice urgently. The FSA was keen to introduce a wide-scale testing regime to assess the extent of the fraud and to combine this with an EU-wide campaign. He asked what tests I thought should be employed and how they might be interpreted. Of the available methodologies, I preferred the 'gold standard' real time polymerase chain reaction (PCR) test for sequencing of mitochondrial DNA, which facilitated a semi-quantitative analysis for horse and many other species; a forensic test which can detect incredibly low levels of different meat species. I felt that my recommendation was perhaps a little ironic as a few years earlier the FSA had funded labs like mine to develop this test, which was not only specific to horse but could be used to identify DNA in all sorts of potential food fraud. Due to insufficient samples, we ceased to maintain the test and our accreditation (vitally important if we were to be involved in prosecutions). Consequently, I didn't have a vested interest in my recommendation; the work would go to another lab which had maintained the accreditation *etc.*

In my opinion, the test would provide reliable and robust monitoring which could be presented in court and would easily detect substitution. The problem with a forensic test such as this is that it would detect miniscule levels of contamination; for example, traces of horse blood left on a knife which had been used in a slaughterhouse to cut horse and then used on a beef carcass. This means we also had to agree a level of horsemeat in a sample which would suggest a benefit to the fraudster, *i.e.* make it worth their while, that might demonstrate a deliberate act for the intent of defrauding or at least making a profit. I therefore suggested to my FSA colleague that we should use the same levels as those prescribed in the Genetic Foods regulations which state 0.9% adventitious contamination.[9] At, or above, this level of contamination, a profit was going to be made by fraudsters (remembering we didn't know the levels of contamination at this stage), and this would rule out accidental contamination, *i.e.* the contaminated knife which would not yield a profit but may show poor hygiene.

To check my pre-dinner musings, I suggested that research be undertaken by someone independent, such as the referee analyst Michael Walker, to check that these assumptions were reasonable. This was subsequently commissioned and published a while later and confirmed my thinking. The conversation finished, and I was called to the table to carve the meat. Yes, you've guessed, that day we served beef, or at least I believed it was?

In an unprecedented step, following our discussion, the UK regulator (the FSA) demanded that all beef products were tested for horse DNA to ensure authenticity, and this was followed across the EU with other agencies asking for the same response.[10]

1.5 THE GROUND-BREAKING PROSECUTION CASES

There were three cases as a result of the investigations, all ground-breaking in their own right. They necessitated a complete change in the manner in which food cases had been prosecuted before, with new guidance for judges, the formation of the food crime unit, a new team at the FSA (adulteration had been elevated to a 'crime') and new working arrangements between food enforcement, the FSA, The Metropolitan Police and the Crown Prosecution Service. All introduced at record speed – four short years between discovery and sentencing.

1.5.1 First UK Prosecutions: Boddy and Moss – Southwark Crown Court March 2015

Following the 'horsemeat scandal' in January 2013, the FSA in the UK began to conduct food traceability exercises at abattoirs known to slaughter horses. Food Business Operators have a legal requirement to be able to identify from which company they have bought products and to which company they have sold them on. This is known as 'one step back' and 'one step forward' traceability. The purpose of the traceability requirements is to ensure that consumers can identify the source of their food and to ensure that unsafe food can be recalled promptly.

Within weeks of horsemeat first being discovered in Tesco beefburgers, on 6 February 2013, West Yorkshire police and inspectors from the FSA conducted an unannounced visit at Peter Boddy Slaughterers in Todmorden, West Yorkshire, as the company had been granted full FSA approval to slaughter horses.

Peter Boddy was the owner and registered Food Business Operator. David Moss was the Manager.

Production at the slaughterhouse was suspended amid investigations into claims it passed off horsemeat as beef for processed foods. The inspector asked to see traceability records for all of the horses that had been processed on site. Initial documents provided by Moss were described in court as 'inadequate' and did not show where a number of horses had gone or where they had come from. However, on that occasion, Moss conceded that some of the horsemeat had been sold 'cash in hand'. He provided the details of one horsemeat customer only: Farmbox Meats in Aberystwyth. The FSA then asked for delivery notes or invoices, and a few days later Moss provided a forged invoice.

Boddy did not produce adequate paperwork for 55 horse carcasses he had sold and accepted 17 animals into his abattoir without the appropriate documents. When questioned, he told officials that 37 had been sold to two Italian restaurants in Manchester and Leeds, but when inquiries were made of the relevant councils, no such businesses were found.

Documents produced were, no doubt, intended to deceive the inspector, and the lack of proper records meant that the source and destination of the horsemeat were untraceable. As a result, food could not be traced or recalled, and it was impossible to know if the meat had entered the human food chain.

The media had a field day providing colourful details such as the fact Boddy was fondly known locally as Greengrass, named after the colourful character in the ITV drama series *Heartbeat*. Boddy often had a range of unusual animals on his premises alongside the more usual livestock, animals such as wallabies, ostriches and even a porcupine. *'He has all sorts up there. People sometimes slow down to look as they drive by because it can be quite a sight,'* said one butcher. *'But he also likes to pull people's legs; he tells them he has an elephant and a silverback gorilla, and in a way he does – big plastic models that he has bought to put in the field!'*[11]

He certainly emulated the 'loveable rogue Greengrass' when he claimed he was innocent. *'I am seriously fed up. This is totally wrong and I will be going to see my solicitor this morning ... I have done nothing wrong.'*

It seems Mr Boddy was no stranger to controversy as, in 2007, he was embroiled in a row when he helped catch and sedate seven cows that had gone into a nearby wheat field. Government vets later told the farmer to kill the animals as they had been sedated with drugs that were not safe to enter the food chain. At the time, Boddy said he had done nothing wrong insisting the *'drugs were perfectly legal'* and adding that the cattle were for breeding not human consumption.[11]

Despite the bonhomie and protestations of innocence, they eventually pleaded guilty and were the first two to be sentenced for passing off horse as beef in an English court since the scandal broke. Clearly, the main problem was the abysmal failure to keep records and track horse carcasses from 'pasture to plate'.

Prosecutor Adam Payter told London's Southwark Crown Court the controversy had *'undermined confidence in the meat industry'*. He added that, although the prosecution could not say whether the meat had ended up in the human food chain, *'there would be no cause to cover up the provenance or destination of horsemeat unless there was something underlying wrong with that horsemeat.'*

Christopher James, defending Boddy, said his client had been reckless but that his failings were not deliberate and the meat had only ever been offered as horsemeat, so there was little or no risk to the public health.[12]

Sentencing them, Judge Alistair McCreath said:

'The traceability of food products, here meat, is of critical importance in relation to public health. If meat causes ill health then it is important that those responsible for investigating the cause of it should quickly be able to discover where the meat came from and trace it backwards to find where the problem lies and prevent the problem escalating.'

Moss was convicted of one count of forgery and received a four-month prison sentence, suspended for two years, and ordered to pay costs of £10 442 within six months.

Boddy was convicted of:

- one count relating to *forward* traceability contrary to Regulation 4 General Food Regulations 2004;
- one count relating to *backward* traceability contrary to Regulation 4 General Food Regulations 2004.

He was fined £4000 per count (total £8000) and ordered to pay costs of £10 442 within six months.

In the 1980s, we seemed to like 'loveable rogues' but I am not sure the judge was enamoured with 'Greengrass'. I got the feeling he had seen a few loveable rogues in his time and, like the rest of us, wasn't keen to accept their lack of compliance with the law.

I suspect this case enabled the newly formed teams to get their act together so that they were 'match fit' for the larger prosecutions to come. I expect 'Greengrass' would say he was simply in the wrong place at the wrong time.

1.5.2 The Second Case: Raw-Rees and Patterson – Southwark Crown Court May 2015

The raid in West Yorkshire led FSA inspectors directly to Farmbox Meats Ltd in Aberystwyth – a meat-cutting plant owned by Dafydd Raw-Rees and managed by Colin Patterson. At the premises, and at an off-site storage facility, they held large amounts of goat meat slaughtered abroad that had been sealed, packaged and labelled ready for sale. However, the meat was falsely labelled as lamb meat or misleadingly labelled as mutton which had been slaughtered in the UK – in short, they dressed up goat from abroad as British lamb.

Although the meat involved was not horse, once again it revolved around false food labels and the traceability of the food that we eat. As prosecution counsel later said in court, '...*the golden rule being that food should be traceable from the farmyard to the fork*', sometimes referred to as 'pasture to plate'.

Investigators found that records at Farmbox were a shambles. Not only were the available invoices insufficient but there were no invoices whatsoever for a large number of products. In addition, there were a 'vast number' of goat carcasses that Farmbox recorded as entering the premises for which there were no records whatsoever to demonstrate who they were sold to and thus where they went. FSA inspectors later found a cache of meat that appeared ready for sale, and of the 19 boxes, 14 were labelled as lamb and two were labelled as mutton. Labels on the goat meat said it had been slaughtered by a lamb abattoir when in fact it came from a company which was not approved to slaughter animals in the UK.

In the police interview, Patterson said it was *'pot luck'* whether a carcass was recorded as goat or sheep as they appeared identical, even to a trained butcher. He claimed he couldn't tell the difference between a sheep and a goat because *'they're exactly the same'*. Raw-Rees told police he had *'limited day to day contact with the company'* and his role was a *'silent partner/owner'*.

Although they both initially pleaded not guilty and the case was set down for trial, Patterson pleaded guilty to 17 counts of falsely describing food contrary to the Food Safety Act and one count of failing to comply with food traceability requirements. Raw-Rees pleaded guilty to one count of failing to comply with food traceability requirements and one count of falsely describing food.

His Honour Justice Alistair McCreath (the same judge as in the case above) described the firm's record-keeping as *'extremely negligent'*. Once again, the judge stressed that traceability matters because it is a vital component in the maintenance of public health given that, if something goes wrong with a food product, it is critically important those investigating can trace it backwards in order to discover where the problem lies and eradicate it. *'This is not a bureaucratic task imposed on business, it really matters.'*

In relation to Raw-Rees, the judge accepted that in the aftermath his business fell apart, the defendant was ruined and his health had suffered. The court heard the company lost £1 million following the investigation and it had since gone out of business. As a result, he was given a conditional discharge for two years.

Addressing Patterson, the judge took a very different approach saying, *'You were the man in real charge of what went on. It was your duty to keep proper records. You did not.'* It was also obvious that the judge took a very dim view of the way he allowed the meat products to be wholly falsely labelled and deprived the public of knowing what they were eating, which also created potential health risks. He was clearly struck by the complete lack of regard of the need to be truthful and, as a result of Patterson's nature and his approach to traceability, Patterson's actions fell within the realms of going to prison. Although the judge imposed a sentence of 12 weeks' imprisonment, suspended for two years, it was quite clear that Patterson came within a cat's whisker of being sent to prison.

This heralded a new era in criminal sentences being imposed on food fraudsters. The signs were not good for anyone who followed. The prosecution team was fully 'match fit'.

1.5.3 The Main Horsemeat Scandal Case

FlexiFoods Ltd – Nielsen and Ostler-Beech
Dino and Sons: Sideras

Ulrich Nielsen, Alex Ostler-Beech and Andronicas Sideras were charged with conspiracy to defraud on the 25 and 26 August 2016 at Southwark Crown Court. Ulrich Nielsen and Alex Ostler-Beech pleaded guilty; Andronicas Sideras entered a not guilty plea.

Between January and October 2012, FlexiFoods (Nielsen and Beech) bought horsemeat and beef from suppliers across the EU and arranged for them to be delivered to Dino and Sons in Tottenham where they would be mixed[13] with the intent of defrauding purchasers by selling goods that contained, wholly or in part, a mixture of beef and horsemeat and selling the mixture as beef. In total, they sold approximately 30 tonnes of meat.[14]

The Italian company had bought the meat from a supplier in Ireland. Paperwork falsified by Mr Sideras was then added as necessary and the meat sold to McAdams.[15] The fake labels suggested meat was delivered directly to McAdams food products from Poland. This was not true, and the Polish supplier stated they do not sell horse; they were the suppliers of beef which was then mixed with horsemeat to make horsebeef. McAdams had the contacts to sell the food to processors who had no interest in buying horse. McAdams and their customers were not 'in on the fraud'. The conspiracy involved Nielsen's Danish firm, FlexiFoods, shipping consignments of horsemeat to Sideras's premises in north London, where they were mixed with beef, repackaged and relabelled as the latter. Beech was Nielsen's 'right-hand man' and worked in FlexiFoods' UK office, where he arranged for the shipments to be transferred and handled the accounting. Unlike the schematic example above, the supplier of the horsemeat was 'in on the act' and knew it was to be mixed or passed off as beef. Emails between the parties demonstrated this knowledge.

Ironically, when you consider the two previous cases, FlexiFoods kept meticulous records, which was the undoing of both companies: FlexiFoods and Dino and Sons. The latter mistakenly delivered the wrong consignments of meat to Silvercrest and Rangelands who rejected the meat, and as a result, this led to the identification of the fraud.[16]

Sideras claimed he was working legally and legitimately held consignments of horsemeat for his business contacts and had relabelled pallets because they were damaged in transit. He was arrested in July 2013 after his fingerprints were found on suspect labels attached to a shipment of what investigators found to be a mix of about 30% horsemeat and 70% beef in Northern Ireland. This load contained the microchips from three horses that had been pets, and their original owners didn't even know they had been sold for slaughter.[16]

The judge, Owen Davies QC, told the men as he passed sentence that the plot was *'not confined to this country, not confined to the firms we have heard about, and it's a big issue for the public to be concerned about, but the fact is it was discovered by accident and only emerged as a problem because of your activity... It's not a mitigating factor, in my judgement, that other people were at it as well as you ...*

The confidence in the food chain was affected adversely, and the share prices of big supermarkets were affected, and it is difficult to recreate the feeling of anxiety that the public had at the time this all emerged.

It was not an activity that was brought to an end by anything other than the arrest of the perpetrators.

The victims in question, properly so-called, of conspiracy to defraud were customers, either wholesalers or the customers of the markets and supermarkets who bought an item that was not what it said it was.'

The prosecuting barrister, Jonathan Polnay, described the conspiracy as a simple process. '*In 2012, beef sold for around €3 [£2.60] a kilogram at wholesale prices. Horsemeat was cheaper. At the time, it sold for around €2 [£1.75] a kilogram.*' The money was made by selling the mix as 100% beef. The majority of the meat, including some from farm horses not sold for slaughter, made it into the food chain and, while the face value of the fraud was £177 869, police said the true cost had probably run into millions of pounds.

Each of the men was convicted of a single count of conspiracy to defraud between 1 January and 30 November 2012.[17]

Andronicos Sideras, 55, was sentenced to four and a half years, Ulrich Nielsen, 58, was jailed for three and a half years and Alex Beech, 44, was given a suspended sentence.

1.6 THE LUCK OF THE IRISH

As reported in the *Irish Times* in May 2013,[8] Ray Ellard, Director of the FSAI said:

> 'There was a massive international food fraud going on and Ireland found it and that's annoyed a lot of people, that the Paddies found it. I'm sorry to say that, but that's the truth of it.'

He said it was felt, particularly in Britain, that the authority was acting on information, but it was just doing its job.

You might suggest that the FSAI got lucky but then isn't luck involved in detecting any fraud? There was more than luck involved as clearly the Irish had the presence of mind to test for horse. I, for one, am disappointed that we didn't make the discovery in the UK, but I don't begrudge colleagues, and indeed friends, in Ireland for finding it, I congratulate them. Maybe it will be our turn next time. It's human nature to want to find something, certainly amongst enforcement professionals. I think the accusation from some at the time was that the Irish were slow in sharing the information; this was perhaps a little harsh as we would all want to check on such a major find before going public.

1.7 IS THERE ANOTHER CASE YET TO BE BROUGHT TO AN EU COURT IN THE FUTURE?

In March 2013, the Department of Agriculture, Food and the Marine in Ireland published a report setting out the findings of an official investigation a mere two months after being informed by the FSAI of its finding of 29% equine DNA in a single beefburger sold in Tesco and manufactured by the Silvercrest plant in County Monaghan. Initially, the investigation involved the FSAI and the Department's veterinary inspectorate and audit teams, but as more facts emerged, it was quickly broadened to include the Department's special investigation unit and the

Garda National Bureau of Criminal Investigations. What had been uncovered was a pan-European problem of fraudulent mislabelling of certain beef products as almost all member states had been affected by the problem. It became a global problem affecting some large global companies and international food brands.

The report *Equine DNA & Mislabelling of Processed Beef Investigation* stated (*inter alia*):

> 'That is not to suggest that intermediaries in the supply chain were the sole cause of the problem. The investigation has also shown direct trade with Poland. In the case of one Polish company, whose product was found positive for equine DNA, the Polish company arranged to collect the consignment and reimburse the Irish operator (QK Meats).
>
> The investigation concludes that in the case of Silvercrest & Rangeland Meats there was no evidence that they deliberately purchased or used horsemeat in their production processes or that these companies were relabelling or tampering with inward consignments.'[18]

It is clear to see that a lot of companies were duped by the trade in horsemeat, and it would not surprise me at all if there may be more cases to come. Headlines such as *'Horsemeat scandal: 66 arrested, with the main ringleader a Dutchman, arrested in Belgium'*[19,20] would suggest there is more to come. Europol said on Monday 17 July 2017 that 66 people had been arrested for trading horsemeat unfit for human consumption and it had seized bank accounts, properties and luxury cars following an investigation into a food scandal that shocked European consumers.

More recently, Spanish police began investigating a group which allegedly slaughtered Spanish and Portuguese horses too old, or in too poor a condition, for human consumption, who forged documentation and sent the horses to Belgium to a large horsemeat exporter.

Is this the end of Horsegate? Food scandals go quiet and the issue isn't repeated for a while as perpetrators think the authorities are looking for this particular issue, so they lie dormant or go elsewhere. Will this be the case for horse substitution? Horsegate certainly heralded a new dawn in public protection; all three prosecution cases resulted in custodial sentences and it brought about massive change in the manner in which the authorities work. This book details the circumstances around this and the creation of a new dawn which I hope moves food enforcement onto a stronger foundation.

REFERENCES

1. Definition of Fraud, The Free Dictionary by Farlex, http://legal-dictionary.thefreedictionary.com/Fraud, [accessed August 2018].
2. Police in UK and Ireland asked to probe more horse meat, https://www.foodmanufacture.co.uk/Article/2013/02/06/Horsegate-latest-police-asked-to-investigate, [accessed August 2018].
3. Victor the horse at centre of horsemeat scandal is 'alive and well', businessman on trial claims, http://www.telegraph.co.uk/news/2017/07/25/victor-horse-centre-horsemeat-scandal-alive-businessman-trial/amp/, [accessed August 2018].
4. Joint Committee on Argiculture, Food and the Marine Debate – Tuesday 5 Feb 2013: Burger Content Investigations: Discussion, https://www.oireachtas.ie/en/debates/debate/joint_committee_on_agriculture_food_and_the_marine/2013-02-05/3/?highlight%5B0%5D=alan&highlight%5B1%5D=reilly&highlight%5B2%5D=chief&highlight%5B3%5D=executive&highlight%5B4%5D=food&highlight%5B5%5D=safety&highlight%5B6%5D=authority&highlight%5B7%5D=ireland&highlight%5B8%5D=food&highlight%5B9%5D=safety&highlight%5B10%5D=authority&highlight%5B11%5D=ireland&highlight%5B12%5D=chief&highlight%5B13%5D=food&highlight%5B14%5D=professor&highlight%5B15%5D=present&highlight%5B16%5D=chief&highlight%5B17%5D=professor&highlight%5B18%5D=reilly, [accessed September 2018].
5. 2013 horse meat scandal, https://en.wikipedia.org/wiki/2013_horse_meat_scandal, [accessed August 2018].
6. Horse DNA found in beef burgers, https://www.rte.ie/news/2013/0115/362894-horse-meat-beef-burgers/, [accessed August 2018].

7. How the horsemeat scandal unfolded – timeline, https://www.theguardian.com/world/2013/feb/08/how-horsemeat-scandal-unfolded-timeline, [accessed August 2018].
8. Equine DNA & Mislabelling of Processed Beef Investigation, Report March 2013, Department of Agriculture, Food and the Marine, http://www.agriculture.gov.ie/media/migration/publications/2013/EquineDNAreportMarch2013190313.pdf, [accessed August 2018].
9. Regulation (EC) No 1829/2003 of the European Parliament and of the Council of 22 September 2003 on genetically modified food and feed, http://data.europa.eu/eli/reg/2003/1829/oj [accessed August 2018].
10. What's behind the horsemeat contamination scandal?, http://edition.cnn.com/2013/02/12/world/europe/horsemeat-contamination-qanda/index.html, [accessed August 2018].
11. Horsemeat scandal: Owner of Yorkshire abattoir denies wrongdoing, https://www.theguardian.com/uk/2013/feb/13/horsemeat-abbatoir-boddy-denies-wrongdoing, [accessed August 2018].
12. Horsemeat scandal: Slaughterhouse boss fined over records, http://www.bbc.co.uk/news/uk-england-leeds-32023036, [accessed August 2018].
13. Businessmen jailed for passing nearly 30 tonnes of horsemeat off as beef, http://metro.co.uk/2017/07/31/businessmen-jailed-for-passing-nearly-30-tonnes-of-horsemeat-off-as-beef-6819388/, [accessed August 2018].
14. Meat supplier 'secretly mixed HORSE meat with beef before flogging it to unsuspecting manufacturers of ready meals and burgers for Tesco and Asada', http://www.dailymail.co.uk/news/article-4671284/Businessman-secretly-mixed-horsemeat-beef.html, [accessed August 2018].
15. Horsemeat plot exposed by equine ID chips in beef, court told, http://www.bbc.co.uk/news/uk-england-london-40559740, [accessed August 2018].
16. Two men jailed in UK for horsemeat conspiracy, https://www.theguardian.com/uk-news/2017/jul/31/two-men-jailed-in-uk-for-horsemeat-conspiracy, [accessed August 2018].
17. Annoyance that 'the Paddies' uncovered horse meat scandal, conference told, https://www.irishtimes.com/sport/soccer/

annoyance-that-the-paddies-uncovered-horse-meat-scandal-conference-told-1.1404808, [accessed August 2018].
18. Europol says 66 arrested in horsemeat scandal investigation, http://www.independent.co.uk/news/business/news/europol-says-66-arrested-horsemeat-scandal-investigation-a7844546.html, [accessed August 2018].
19. Dozens arrested in Spain for Europe-wide horsemeat scam, http://edition.cnn.com/2017/07/17/europe/european-horsemeat-ring-busted/index.html, [accessed August 2018].
20. 66 people arrested in Spanish horse meat investigation, https://www.rte.ie/news/2017/0716/890615-horsemeat/, [accessed August 2018].

CHAPTER 2

The Grecian Temple of Food Enforcement

The public, led by the media, had previously shown only a fleeting interest in food adulteration (fraud), even when serious health effects were noted. Sudan Red (2005), the chicken scandal (1990s–2008) and melamine in Chinese baby milk (2008) all received some media attention. Sudan Red, the carcinogenic food dye, resulted in one of the largest recalls ever, the 'pumped-up' chicken scandal was EU-wide and evaded enforcement for many years and melamine in baby milk killed six infants, hospitalised 50 000 and affected 300 000, but surprisingly no lasting legacy or media attention followed any of these scandals, just like many other scandals that were seen before. Perhaps the only issue previously noted that led to major change was the Bovine Spongiform Encephalopathy (BSE) in cows, which led to a government reorganisation and the introduction of the FSA. After each previous major scandal, a professor was asked to carry out a review on behalf of the FSA; each time they published a report and that was received by the FSA and discussed at board level. No lasting legacy resulted and no change in practice, at least not to anyone observing from outside of the FSA. Many of the hardened enforcement professionals (including me) thought that horsemeat was just another issue, here today and barely remembered

---

The Horse Who Came to Dinner: The First Criminal Case of Food Fraud
By Glenn Taylor
© Glenn Taylor 2019
Published by the Royal Society of Chemistry, www.rsc.org

after a short time-frame. It seems we were very wrong. Somehow this is different, at least so far. Is that due to a love of horses and a hatred of eating horsemeat amongst the English? Was it due to the review carried out by Professor Elliott? The evidence suggests that a legacy will remain; one that will change food enforcement for ever.

Since 1860, UK food prosecutions have been based on the strict liability crime of food adulteration, not fraud. The term food fraud was not used by food enforcement or the regulator, probably as it is far more difficult to prove. Fraud prosecutions require the prosecution to establish the intent (*Mens Rea*) as well as the act (*Actus Reus*). The act relies on an analysis to prove the food was not of the nature, substance nor quality outlined in regulation; difficult to dispute and relatively easy to prove. The UK was out of step with its EU partners who had pursued crimes such as food fraud for around 60 years. Criminal offences such as fraud attract larger penalties than statutory offences; more often than not, a Magistrates Court will fine a defendant who is found guilty of a statutory offence whereas those found guilty of fraud in a criminal trial may well receive a custodial sentence.

2.1 EIGHT PILLARS TO SUPPORT THE FOOD ENFORCEMENT TEMPLE

Professor Chris Elliott, Professor of Food Safety and Director of the Institute for Global Food Security at Queen's University Belfast, was asked by the Secretaries of State for DEFRA (Department for Environment, Food and Rural Affairs) and Health to carry out an independent review of Britain's food system in the light of the 2013 horsemeat fraud.[1] This in itself was not unusual as, each time a major scandal had been noted, a full review was called for by the regulatory bodies and undertaken by an academic of similar rank, but this one was different. Professor Elliott consulted far and wide and produced a report which stayed in the public eye for longer than usual. Perhaps it is just a coincidence but Professor Elliott's report suggested eight pillars to underpin food enforcement and ensure its stability, free from fraud; the Grecian Temple of Food Enforcement? Eight pillar temples such as those built by the Greeks can be matched

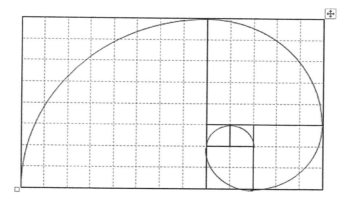

Figure 2.1 The Fibonacci spiral, or golden spiral, demonstrating the golden ratio, wherein the width of the spiral increases logarithmically in relation to its point of origin by each quarter turn.

to a Fibonacci spiral (Figures 2.1 and 2.2(A) and (B)). The spiral follows a 1.1618 ratio, which occurs naturally (Figure 2.3) and has been considered throughout history as pleasing to the eye.

Greek temples were built to demonstrate power and stability of the Greek reign. Remove one pillar and the structure is no longer as strong and much more prone to collapse and no longer a beautiful structure. Was this just coincidence? It certainly was intended to show power in response to fraudsters and to continue the fight against fraud for a long time. Media attention in this review and the outcomes has helped maintain interest, and Professor Elliott had a masterstroke up his sleeve: he challenged the government to introduce *all* of the eight pillars of food integrity reform that he identified, not cherry pick one or two, as he felt it necessary to adapt all to maintain food that is free from fraud. I am sure those responsible were desperate to cherry pick; who wouldn't be? Carry out the quick easy wins and shelve the others. For me, that was a thing of beauty as far as this review was concerned – it was not possible to cherry pick, not this time; Chris Elliott was adamant that the government wouldn't try this, and to ensure they didn't, he reviewed the response of government to his report, publically publishing the results. The section below examines the performance of enforcement and regulators against one of Professor Elliott's reviews of the progress made.

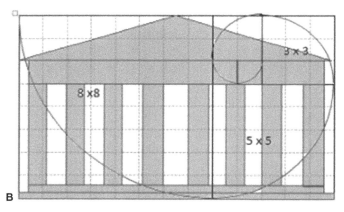

Figure 2.2 (A) The acropolis of Athens, an example of an eight pillared Greek temple (© Shutterstock). (B) A schematic showing the spiral overlaid on a Greek temple, demonstrating the construction to reflect the natural aesthetic and strength of the Fibonacci ratio.

2.2 SO WHAT WERE THE EIGHT PILLARS AND WHAT PROGRESS HAS BEEN MADE SINCE HIS REPORT?

2.2.1 Pillar One: Consumer First

A team led by Stephanie Brooks wrote a review: 'Four years post-Horsegate: an update of measures and actions put in place following the horsemeat incident of 2013'.[2] For each of the eight

Figure 2.3 The aesthetic effect of the Fibonacci spiral in the structure of a nautilus shell (© Shutterstock).

pillars, I shall quote from this report and add my own commentary afterwards.

> 'In the weeks following the horsemeat incident, consumer trust in beef products decreased to an all-time low. Many media and research institutions including Which, The Consumer Council and Kantar Worldpanel reported decreases in UK consumer trust in processed meat products and in food retail outlets themselves. Kantar Worldpanel reported a 43% and 13% drop in frozen burger and frozen ready-meals sales, respectively, in the 4 weeks leading up to 17 February 2013 compared to the same period in 2012. Consumers' shopping habits changed as a result of Horsegate with a consumer survey revealing 7% of consumers stopped purchasing meat altogether and 65% of consumers trusted food labels less as a result of the incident. A study into consumer confidence post Horsegate found that consumers expressed a sense of betrayal and concern over the complexity of supply chains. A Consumer Council report published in July 2013 reported that almost half (47%) of consumers saw food retailers in a less favourable light in the months following the incident, and a 24% decrease overall in consumer trust in the food industry was reported in March 2013.

The Elliott Review recommended that consumer needs be of the highest priority for industry in order to ensure consumer confidence in the food they purchase. It was recommended that committing food crime should be made as difficult as possible through the prevention of contamination, adulteration and false claims. Which *(the consumer magazine) had consumers interests at the forefront when they began their campaign entitled "Stop Food Fraud" calling on government, the FSA and local authorities to help stop food fraud. This campaign received 37,000 consumer signatures and is regarded as one of the integral initiatives that helped rebuild consumer trust in the food industry following Horsegate.*

Since Horsegate, Professor Elliott believes food retailers' attitudes have changed drastically, with transparency now a key trend. For example, Tesco provide information online on their meat product testing regime and is widely regarded as a mechanism for trust-building among consumers. Following the campaigns and initiatives put in place by the industry, an improvement in consumer trust levels in the meat industry and supermarkets was reported, with consumers' trust levels returning close to pre-Horsegate levels. It is believed these campaigns and initiatives taken to put consumers first have helped the industry regain consumer trust.'

Back in 2013, Findus said: '*At Findus UK our first priority is our customers and providing quality products they can trust. But we know that many people have been concerned by the news this week that tests have shown that some of our Findus Beef Lasagne has been found to contain horsemeat. We understand those concerns, we are sorry that we have let people down and we want to outline the facts.*[3] *The Findus name has been withdrawn and its investors have decided to re-brand the products.*'[4,5]

I think all food businesses would state that their first priority is to the consumer. At the time, consumer confidence in food retailers and businesses was at an all-time low, and how this compared with previous issues is an interesting research topic. I postulate (without much evidence) that there is a cyclical wave of mistrust each time a food scandal rears its ugly head. Some scandals will have more impact than others but this is probably directly proportional to the amount of media coverage and how close the scandal is to home and what risk to health. The melamine issue may have been seen as an issue for the Chinese, for example, and I was surprised it wasn't more of an issue at the time. The water in chicken fraud should have had a similar impact to Horsegate, but it didn't achieve the same media coverage. Horsegate was of far more interest to the media. I was guilty at the time of expecting Horsegate to blow over quickly as previous scandals had done in the past. Indeed, this was a comment that many of my colleagues in enforcement made. Horsegate was different, perhaps because horses are pets and the British adore their pet horses and dogs. Indeed, enforcement colleagues from across the EU ribbed me regarding the British reaction to Horsegate; they watched with amusement at our reaction. How wrong we were, clearly Horsegate was different.

Communication quickly became an issue. The FSA changed its strategy and others quickly enhanced their communications with the public. I am not sure that this was as a result of Professor Elliott's report as it happened during the scandal, before his report was published. I thought one company's Chief Executive made some interesting comments at the time and that these might have damaged the reputation of his company. He made me laugh with his comments about testing for hedgehog. He was forthright, and it turns out he correctly read his customer base and I was wrong, no damage being done to the reputation of that company.

I am pleased to read that food businesses want to publish their test results and share them with consumers, but communication is two way and I am not sure how much attention is paid by consumers to day-to-day communications about what testing has been undertaken and the fact that the results were satisfactory. Isn't that the expectation of the consumer: that food is risk-free and of the nature, substance and quality they demand?

To be fair to all the major food businesses, they were putting their customers first. They interpreted what they thought their customers want and they did their best to provide it whilst making sufficient profit to satisfy investors. The mantra of the FSA had, since its inception, been a version of putting the consumer first.

So, in summary, I'm not sure there has been a massive change in terms of putting the consumer first, nor has this resulted in an improvement in attitudes of the consumer towards food businesses. Consumers have an expectation of an honest food industry; if it says beef on the label, they buy in the expectation that it is accurately described and compliant with the law. What we did learn was how important communication is, especially at a time of crisis.

2.2.2 Pillar Two: Zero Tolerance

The review team wrote: *'The Elliott Review recommended the food industry adopt a zero tolerance to food fraud and the food industry was encouraged to question whether some procurement deals were "too good to be true", i.e. procurement of raw material that is suspiciously inexpensive in relation to anticipated cost of production, to ensure due diligence in procuring safe and genuine food from reputable sources. Further, food crime/fraud and mitigation strategies were to be considered as part of risk assessment procedures and whistleblowing/reporting of food fraud and crime was strongly encouraged. A whistleblowing hotline has been subsequently set up by the National Food Crime Unit (NFCU),* (see pillar 7). *Additionally, it was suggested that industry should be rewarded for responsible procurement and sourcing practices and the review encouraged industry to carry out sampling and testing within their supply chains.*

Previously within the meat industry, "a little bit of cheating" would have been considered acceptable within the moral boundaries of "normal practice". For example, Horsegate can be seen as a consequence of this "little bit of cheating" where Polish beef was procured, packaged and labelled as British beef. While the food company believed they were operating inside the moral boundaries of "normal practice", the consequences of these practices came when horsemeat was found within meat procured from Poland and sold as British beef. It is a belief that the industries are now more aware that they are being monitored extensively by their customers through the rigorous testing programmes implemented since Horsegate and as a result attitudes to cheating have changed. However, a recent online publication suggests illegitimate actors (fraudsters) are continuing to mastermind and facilitate food fraud. Despite this, it is believed there is much greater awareness of food fraud and a realisation among legitimate meat industry actors, such as processors, that shortcuts cannot be taken regardless of downstream pressures. Genuine food businesses are aware that any deviation from customer specification can have great consequences for their businesses. While this is thought to be the case within the meat industry, this unfortunately may not be the case in other sectors where shortcuts are still considered within "normal practices". This illustrates that there is a continued tolerance towards food fraud by some actors and they may not cease until abuses result in injury or death as with the Melamine incident in 2008'.

This is one of the major successes of Professor Elliott's review. The industry and regulators have now awoken to the possibility of food fraud and not a lesser crime of adulteration. Much more attention is paid to this issue directly as a result of the review. The continuing issue will be how to stop fraud. A sufficient risk of detection, together with punishment which acts as a deterrent, is essential. These necessitate a continued focus by industry and enforcement and, of course, the necessary resources. In January 2018, the FSA announced that the food crime unit would only take on the investigation of food fraud if it received between £4 and £5 million extra funding each year from government. Given that local authorities haven't investigated food fraud and aren't set up to take on this additional responsibility, this new money, if allocated, will demonstrate the government's continued commitment towards fighting food fraud. The effects of this food scandal have lasted longer than nearly all predecessors. Will it continue? I, for one, hope so. The Chancellor's Spring Statement (March 18) has allocated £14 million in additional funds to the FSA, and amongst this is the money required for the food crime unit (subject to a satisfactory business case being provided). This is excellent news and demonstration of further commitment of the government towards fighting food crime and non-compliance.

2.2.3 Pillar Three: Intelligence Gathering

The review team wrote: *'Government and industry focus on intelligence gathering and sharing was the take-home message from recommendation three. In addition, the development of a "safe haven" for industry to collect and disclose information and intelligence (whistleblowing) without fear was recommended. Protection of* whistle-blowers from retaliation has also been highlighted as a necessity by the United States Food Safety Modernisation Act. Previously within the food industry, technical counterparts from different companies informally discussed suspicious supplier behaviours or practices, but this information was never shared with regulators and governmental departments for fear they would be implicated in the suspicious activity. Top legal firms in the UK advised their clients not to share information or intelligence with the regulator unless absolutely necessary.

Since Horsegate, in the UK a system has been established to facilitate intelligence gathering and sharing within industry called the Food Industry Intelligence Network (FIIN), which covers all commodities and about 60% of all foods sold in the UK. The FIIN network allows companies to share anonymised information and test results via a legal firm which ensures that competitive advantages cannot be gained from one company knowing another's results. The anonymised information is passed on to the FIIN network for analysis and subsequently shared with member companies. A lot of company resource has been channelled into testing thousands of meat products in light of Horsegate and results indicate there to be no detectable fraudulent activity in meat tested at the current time, but this does not suggest food fraud is gone for good. Professor Elliott explains there is a distinct need for product testing to be dynamic to ensure other commodities are examined sufficiently. Despite the significant steps the industry has taken in sharing information and intelligence with each other, information is still not shared with the regulator or government departments as recommended by The Elliott Review, due to issues surrounding confidentiality, competition and fear of implication. Due to the government's commitment to a high level of transparency with the public, food businesses have concerns regarding government's (FSA) intentions in sharing of highly sensitive company information such as company name. Food businesses want assurances that personal company information will not be shared publically. As this written guarantee has not yet been provided, the FIIN network information remains unshared with regulators and government departments. Interestingly, Food Standards Scotland (FSS) has agreed to provide this written guarantee and thus FIIN will share their information with FSS but not the FSA at the present time. The effect this may have on a potential shift in the FSA's position remains to be seen.

While not solely a UK initiative, The Food Fraud Network (FFN) is another intelligence gathering initiative established post-Horsegate. The FFN was set up in 2013 and comprises 28 national food fraud contact points in a European regulatory sharing network, working closely with INTERpol and Europol on Operation Opson, targeting organised food crime networks. However, this information is not shared with any other party (including industry). Intelligence gathering and sharing was recommended and intended to be a joint activity between regulators and industry. However, this appears not to be the case as industry and regulators are actively gathering and sharing information and intelligence within their respective groups but not with each other'.

Again, another excellent initiative and one that regulators must do all they can to embrace. FSA and other government departments need to find ways around revealing information they hold on file that has been gathered in this way. The concern at the time was freedom of information queries leading to disclosure of all information. It is intelligence, so I never did understand why a guarantee couldn't be given. I realise that FSA want to maintain their position of trust and independence as representatives of the consumer and in the eyes of the public, and I am heartened to hear that FSA Scotland has managed to find a way around these issues. Before retirement, I worked on an intelligence group with a team from the FSA, other regulators and representatives of the large food businesses and trade associations, and I can vouch for the usefulness of open dialogue. We did not share results, just concerns and issues and how to detect them, what the information means and how to manage the issue. This was incredibly helpful to all members. Anecdotal information was just as useful as that backed by tests. More work is needed here, but a great start. The work with EU partners as part of Opson has given good results and increased effectiveness of the food crime unit.

2.2.4 Pillar Four: Laboratory Services

The review team reported: '*The horsemeat scandal revealed that laboratory services for food fraud testing were not up to the required standard. The quality of testing was unknown and the number of laboratories that could provide such services was decreasing due to a downturn in government spending. In light of this, the Elliott Review recommended that access to resilient and sustainable laboratory services with standardised and validated testing methods be implemented as a priority. It urged the government to facilitate the development of surveillance programmes (intelligence gathering, horizon scanning) and targeted sampling programmes based on intelligence gained'.*

The final Elliott review report stated (*inter alia*): '*public analyst laboratories are in a fragile position and this review provides an opportunity to develop a sustainable national asset. . ..*

It is government's responsibility to provide resilience in the event of an emergency and to benchmark for standards of testing. . ..

Continuing technical development that supports food authenticity and safety testing requires a healthy mix of private and public sector laboratory provision. The best way of securing this is to bring together the remaining public sector laboratories into a merged, shared service. A public sector "spine" to laboratory provision for food testing would create a resilient, competitive service that would prevent laboratory provision becoming monopolistic. The creation of a modernised, integrated, national "Local Authority Laboratory Service" comprising Official Food and Feed Control and other public health protection would offer considerable added value. This would lead to the Service working alongside private sector provision. Private sector delivery in the Official Control Laboratories (OCL) system has demonstrated that through establishing a "critical mass" of public analysts operating as a team, optimised utilisation of equipment and buildings and "lean" management, efficiency gains are possible and many technique areas are scalable. The current local authority provision could benefit from developing similar efficiencies. Bringing together the remaining six public sector laboratories in England, through a process of strategic rationalisation, would secure a scientific service that is accustomed to dealing with local issues and understands their context, but supports nationally planned programmes. This would create a sustainable national asset, comparable with PHE's microbiological laboratory network and there is a significant degree of common ground between potential partners when it comes to investments in plant, equipment, and staff. . ..

In addition to their central role in relation to food and feed law enforcement, public analysts also provide expert scientific support with a broad public health focus to local authorities and other bodies in a wide variety of areas. Whilst each laboratory has its own individual mix of analytical work they undertake, they all agree that none of the remaining public sector laboratories would survive without the additional revenue provided by these analyses. The possibility of absorbing these activities into a shared service could also be considered.

A shared, merged public sector laboratory service is the only option to secure the public sector laboratory system. This sort of project has been discussed before, but has never progressed. The urgency is increasing; in 2010 there were 10 public sector public analyst laboratories, now there are only six (this number has reduced to four, May 2018). *This may be the last opportunity to create a resilient, robust, shared service that will provide a sustainable national asset to the UK. It needs to retain momentum, and the commitment of those involved. There are two key roles here; one to provide objective, neutral facilitation of the project group and one to scrutinise progress and make recommendations for how the process can be improved. The first role could be fulfilled by a professional body like the Institute of Food Science and Technology and the second by the House of Lords Science and Technology Select Committee.*

He concluded: *Recommendation 4 – Laboratory Services: Those involved with audit, inspection and enforcement must have access to resilient, sustainable laboratory services that use standardised, validated approaches.*

The government should:

- *Facilitate work to standardise the approaches used by the laboratory community testing for food authenticity;*
- *Work with interested parties to develop "Centres of Excellence", creating a framework for sandardising authenticity testing;*
- *Facilitate the development of guidance on surveillance programmes to inform national sampling programmes;*
- *Foster partnership working across those public sector organisations currently undertaking food surveillance and testing including regular comparison and rationalisation of food surveillance;*
- *Work in partnership with Public Health England and local authorities with their own laboratories to consider appropriate*

options for an integrated shared scientific service around food standards; and
- *Ensure this project is subject to appropriate public scrutiny'.*

I must declare an interest. I am not a public analyst, but at the time, I was responsible for one of the six (now four) public analyst services. I had seen the number of labs reduce from around 40 to the four or so 'in business' today. Like many, I felt that this was tragic and desperately sad for a profession which had served the public so well over the last 150 years or so. The problem was that the profession hadn't scanned its business horizon and they refused to accept the signs that major change was imminent.

- 26 out of 27 EU member states had state sponsored Official Control Laboratories (OCL) – not public analysts.
- The only other member state with an equivalent to a public analyst role was Germany, where the qualification was thriving and producing people to work in enforcement and the food industry.
- One member state had already moved to a model of employing a private sector OCL.
- There were only a handful of qualified public analysts working in the UK, most were rapidly approaching retirement and perhaps one new person gained the qualification each year.
- Some private companies see the qualification as a barrier to entering the food enforcement market, but they work for the food industry providing advice and analytical testing.
- In the 1980s, many public analyst labs gave their budgets to trading standards colleagues (in the same local authority) to 'spend' with the laboratory on food analysis. This increased efficiency initially and staved off compulsory competitive tendering (all the rage at the time amongst local authority colleagues) but led to reductions in spend over time. This led to reductions in development of new techniques and training of future public analysts. A vicious circle which wasn't foreseen had happened and this has led to the demise of the profession.

- As a result of competition, public analysts didn't share expertise nor collaborate as well as EU counterparts. Several attempts were made, but alas without success.
- Our EU colleagues in Germany, in particular, were staggered at our lack of sharing. I was offered a method for food authenticity developed by one of the German labs with a local university. This tested geographical origin of food using isotope ratios. It cost around £250 000 to set up and equip. Our models of funding could not cover that sort of expenditure. In the UK, we would have needed thousands of samples over a relatively short time-frame to persuade our local authorities that we had a suitable business case. As a result, few of us developed new techniques for testing of food.
- The Eurofins model is very successful; they took over public analysts labs, engaged the public analyst and took the analytical work to their specialist labs. They worked for the food industry and, as a result, were able to develop new technology and techniques as the costs were amortised over a wider client base, with the food industry effectively subsidising enforcement.
- The numbers of samples submitted to labs falls year on year and is a small fraction of what it was.
- In the last few years, under the leadership of John Barnes, the FSA provided an annual sum for food enforcement projects. This amounted to around £1 million each year and kept labs viable, and it was just a question of time before the funding stream was removed (it happened when John retired).

The stronger focus on food crime is likely to divert FSA resources (see last bullet point) away from local authorities and they will be left to fend for themselves. Local authorities should be held accountable but, with around 45% of funding reductions for local authorities since 2008,[6] this isn't likely to be a high priority for them. The FSA has once again reiterated that it has no intention of taking responsibility for local authority food enforcement work. Why would they, the budget is not there? In her recent article,[7] Felicity Lawrence lamented: '*Local authority sampling and testing of foods fell by a quarter in 2017.*[8] *At the same time council public analyst laboratories that carry out these tests are disappearing, as their work for depleted trading standards teams dries up. These cuts may seem safe to make. They're not. Will it take a serious outbreak of food-borne illness for the government to realise what we are losing?'*

To ensure the viability of the professional qualification, by creating more people to train, one option was for the public analysts to develop an alliance with others – for example, our German counterparts who were interested in working together to develop a joint model; theirs was similar and very successful. Consequently, I worked with them and our Austrian colleagues who were interested in developing a similar qualification in their member state. Sadly, this failed to gain momentum with UK public analysts.

Perhaps we should look towards the Government Chemist (LGC) as an exemplar. This started in 1842 as a laboratory focussed on regulation of tobacco to ensure it wasn't adulterated and to protect government revenue (tax due from imports of foods *etc.*). A referee function was added under the Sale of Food and Drugs Act 1875 (the Act which also introduced public analysts). This referee function was designed to be independent from the public analyst and any laboratory appointed by food businesses. Where there was a dispute between the analysts, the referee would carry out an independent analysis and adjudicate. The current post holder is Michael Walker. He shares his results with interested parties on an annual summarised basis. It is very rare indeed for him to disagree with the results provided by public analysts. This is testament to the skills of the public analysts. LGC was privatised in 1996 and is currently owned by an investment company. Since privatisation, there has been a 10-fold increase in staff. The company is a national reference laboratory which operates to the highest quality standard and manufactures 'reference materials' for other laboratories to use to monitor and demonstrate their capabilities, either as part of a ring-trial (external proficiency assessment) or in-house check. It is a world leader in many areas of science and is able to utilise the very latest technology.

The few remaining local authorities who provide their own public analyst service have signed an agreement to work together, but very little more has been achieved. Sadly, this recommendation is doomed to fail; perhaps a better model would be, if we do not want a monopoly of provision, to work with another private sector partner to provide an alternative to Eurofins. Both LGC and Eurofins are very highly regarded with excellent pedigrees; perhaps we shouldn't fear the private sector.

2.2.5 Pillar Five: Audit

The review team wrote: *'The Elliott Review highlighted food audits as too food safety focussed and food fraud awareness within auditing to be poor. Consequently, the review recommended developing a modular approach to auditing underpinned by both food safety and integrity standards. The industry has taken this on board and taken several steps to address the recommendation on a wider industry level. The existing British Retail Consortium (BRC) Food Safety Standard (adopted by manufacturers as a pre-requisite for customer supply) has been adapted (Issue 7) to include new mandatory clauses concerning vulnerability assessment. Food manufacturing sites are expected to carry out frequent risk assessments of their products by completing vulnerability assessments on raw materials procured directly (from the manufacturer) or indirectly (via an agent/broker) and establishing mitigating strategies to reduce any identifiable risk. Additionally, where raw materials are procured via an agent/broker, sites must understand and assess the manufacturer's suitability as a supplier. It is important to note that these new clauses apply to all food sectors and not just the meat sector.*

The Elliott Review identified traders and brokers as an area of vulnerability in the chain and several initiatives have been adopted by industry to address this. The BRC Standard for Agents and Brokers was developed to allow reputable agents and brokers to be certified against a standard which ensures they have processes in place to manage their supply systems, with a particular focus on food fraud prevention. BRC Global Standards state that UK food retailers are actively encouraging their agents and brokers who supply retailer Own Label produce to become certified against this standard. The BRC Standard for Agents and Brokers was implemented on a macro industry level and applicable to all agents and brokers (not just meat agents and brokers) and does not take into account the specific nature of the meat industry. The IMTA Good Trading Practice Guide to food fraud prevention in 2015 is tailored specifically to meat traders and brokers and sought to "identify practical steps which companies could take to strengthen their resilience to fraud threats" and is available to the International Meat Traders Association (IMTA) members and non-members on request. In addition to the guide, IMTA have introduced a "meat scam tracker" where members can report suspect scams across the industry. Following active engagement with the NFCU (see pillar 7), IMTA members are seen to be proactively and regularly sharing intelligence with IMTA which is subsequently passed onto the FSA. Since the horsemeat scandal, IMTA have worked to ensure better information sharing with government and acknowledge the efforts of their members to become more food fraud aware via the implementation of preventative measures in their businesses.

In conjunction with the food industry and the British Meat Processors Association, BRC Global Standards have also developed a voluntary bolt-on module to the Food Safety Standard specifically for meat companies called Meat Supply Chain Assurance, which goes back to the point of slaughter. This module enables the ability "...to demonstrate to customers an increased transparency of their meat supply chains..." and is designed to enable meat companies to demonstrate increased traceability and visibility in their supply chain and how they manage species–species contamination. The Meat Supply Chain Assurance bolt-on is believed to have removed some of the additional audits triggered in the immediate aftermath of Horsegate.

The Elliott Review recommended regulators and industry work together to develop an appropriate auditing training platform for food fraud detection. While this has resulted in a market opportunity for auditing bodies/companies, it is still in the early stages of inception. Training of auditors to be forensically food fraud aware is fraught with difficulties as training materials and methods cannot enter the public domain, as this would create an awareness of the auditing scope and criteria thus facilitating fraudulent companies and individuals to commit further food crime/fraud.

The review recommended that the number of audits be decreased through using a risk assessed approach, i.e. giving credit where credit is due, but the quality of these audits should be improved to *ensure food fraud is an integral part. Contrary to other beliefs, Professor Elliott believes the number of audits are thought to have actually increased in light of Horsegate, with individual customers continuing to carry out their own audits. Auditing is a business in itself, where a substantial amount of money is generated, and could be regarded as a conflict of interest to rationalising and reducing the number of audits. However, there have been significant steps taken by food retailers to move to an unannounced platform as recommended by The Elliott Review; 99.5% of Asda audits and more than 50% of Tesco audits are now unannounced, while M&S has created a new unannounced audit platform specifically looking at Food Integrity'.*

This is another success, and a great deal of work has been carried out to develop useful training and accreditation schemes across industry with regulators and food consortia. Companies are far more likely to focus on fraud as part of their audit and to look for the unusual.

2.2.6 Pillar Six: Government Support

The review team wrote: *'Ensuring a joint and inter-disciplinary approach to fighting food fraud between high-level officials in FSA, Department for Environment, Food and Rural Affairs (DEFRA) and Department of Health (DH) was recommended by The Elliott Review. In light of this, the Cross-Government Group on Food Integrity and Food Crime has been established and is chaired by the DEFRA Minister for Food and Farming, George Eustice, and attended by ministers from the DH (Public Health), Home Office (Organised Crime), Business, Innovation & Skills (Consumer Affairs) and the Chair of the FSA. The high-level official group meet bi-annually, with senior officials in these departments providing support through regular meetings. It is the regular collaboration and communication between senior officials, outside of high-level meetings, that creates connectivity between departments'.*

There is evidence of good quality collegiate working, but I have a bias here as I believe food should have one voice and the FSA is best placed to hold the government to account and take action. The FSA was set up as a non-ministerial department to hold government accountable and food enforcement needs that support.

2.2.7 Pillar Seven: Leadership

The review team wrote: '*The nature and magnitude of the Horsegate incident was something not previously encountered by the regulatory authorities, and consequently meat fraud had been overlooked previously. On the other hand, there is widespread awareness of fraud in the olive oil and fish industries. The uniqueness of the situation meant that in the initial days and weeks after the incident broke, there was a lack of clarity in who was to respond to and lead the investigations into the incident—the FSA or DEFRA? This was reaffirmed in the Elliott Review where it was identified that there was no one body dedicated to fighting and preventing food crime/fraud in the UK and as a result, the introduction of a NFCU was recommended. The Elliott Review investigated food crime investigation infrastructures in other countries, namely the Netherlands, Denmark, France and Germany, and concluded the UK system should be modelled on a system similar to that employed by the Dutch which has full police powers and has been established for over 60 years.*

The NFCU, set up in December 2014, was placed within the FSA and is headed by a former senior intelligence officer with experience in several law enforcement agencies including the Serious Fraud Office and National Crime Agency. The role of the NFCU is to "give greater focus to enforcement against food fraud in government by analysing intelligence, initiating investigations and liaising with other criminal and regulatory enforcement agencies" and it leads the response to food crime in England, Wales and Northern Ireland. In June 2016, the NFCU launched Food Crime Confidential, a whistleblowing hotline where food crime can be reported safely and anonymously. While significant steps have been taken to implement the NFCU, it is regarded as significantly under-resourced, with a £900,000 annual budget (as opposed to £2–4 million per year as recommended by The Elliott Review), to carry out any meaningful activities. The new Chair of the FSA has recently commissioned a review of the NFCU to determine its future direction'.

This is critical to successful food fraud prevention. The National Food Crime Unit (NFCU) has just been awarded the funding recommended by Professor Elliott and has made very successful interventions to date and formed good partnerships across the EU working on projects such as Opson. The need to prove fraud will require a suitably trained enforcement workforce which understands how to prove intent and manage unused material (material gathered in an investigation which may help or hinder a defence and needs to be shared – something that some police prosecutions currently struggle to master). The food crime unit has such expertise. It has been suggested that local authorities will continue to investigate statutory offences and that the food crime unit will investigate criminal activity (fraud). Slick working relationships will be needed and a good understanding of roles *etc.*

2.2.8 Pillar Eight: Crisis Management

The review team wrote: *'The Elliott Review recommended that effective mechanisms needed to be implemented in order to sufficiently deal with serious food safety and/or food crime incidents in the future. The review stressed the importance of defining roles and responsibilities in the FSA before another food safety and/or food crime incident surfaces. While there hasn't been another major red meat-related incident since Horsegate, there have been other major food fraud incidents uncovered in other sectors including in the herbs and spices, and honey industries. A study carried out by the Institute for Global Food Security at Queen's University Belfast found 24% of oregano samples purchased from UK retailers and online sources had been adulterated with other non-oregano ingredients. Professor Elliott believes there has been a significant learning in dealing with major food incidents and government is now much better equipped to deal with major food fraud incidents, compared to when Horsegate surfaced in 2013.*

It is believed there is a much greater understanding among industry and government of how food fraud incidents like Horsegate can occur in these complex supply chains. While significant improvements in the understanding of fraud and the application of fraud prevention have been made in industry and government, it is important to note that "fraud hasn't gone away, [and that] fraud [type] just changes". This illustrates that the fight against food fraud/crime is a continuous process as fraudsters continue to evolve and find innovative ways to infiltrate supply chains as long as human greed exists. It is therefore important to ensure mitigation and detection strategies are at least equally innovative and evolutionary.

By mapping the UK beef supply chain, this paper has provided context as to how the nature of complex and convoluted supply chains can create vulnerabilities open to exploitation by opportunistic food fraudsters, particularly within the EU. The Elliott Review's eight pillars of integrity recommended measures to help improve the integrity of food supply systems. Significant steps have been made by both industry and government to implement some of the recommendations in the UK. Industry attitudes to "a little bit of cheating" have changed substantially. Testing and surveillance systems that have been integrated into normal industry practice, and the government are more prepared for future incidents through the establishment of the NFCU and the defining of roles and responsibilities. Horsegate substantially raised the profile of food fraud and crime occurrence within supply chains and, despite improvements to date, further collaboration between industry and government is required in order to align fully with the recommendations set out in The Elliott Review. Ensuring the sharing of intelligence across industry and government and adequate resource allocation are particularly important in the fight against food fraud. Lessons learnt in the UK and Ireland in relation to horsemeat and the subsequent implemented recommendations and initiatives from The Elliott Review are not only applicable to UK and Irish contexts. They are also relevant to other jurisdictions to enable the implementation of safeguards in preventing food fraud and crime in their own country'.

It is correct that Horsegate has led to a focus on food fraud and how to identify it and prevent it and how to communicate with the public. The round table meetings chaired by the FSA during Horsegate were a new initiative which worked very well and showed both industry and fellow regulators how to manage the crisis. This led to an EU-wide initiative to identify the extent of horsemeat fraud.

2.3 WAS IT A SUCCESS?

There is no doubt in my mind that Chris Elliott did a great job with his review, but perhaps he was fortunate in that this scandal involved horses and, as such, gained widespread attention, unlike many before. If the lasting legacy of horsemeat is an improved mechanism and focus on food enforcement involving large food businesses working with a centrally coordinated enforcement/regulatory team which offers a cogent threat to food criminals, then Professor Elliott has succeeded where others failed and he has left a lasting legacy. Early indications suggest that this will be the case. To my mind, this was an exemplar. Challenge government to carry out the whole review and nothing but the whole review, and in that way, a more lasting legacy is likely, and all achieved during a time of severe austerity when the last thing governments want to do is spend more on food regulation and enforcement.

REFERENCES

1. Independent report: Elliott review into the integrity and assurance of food supply networks: interim report, https://www.gov.uk/government/publications/elliott-review-into-the-integrity-and-assurance-of-food-supply-networks-interim-report, [accessed August 2018].
2. Four years post-horsegate: an update of measures and actions put in place following the horsemeat incident of 2013, https://pure.qub.ac.uk/portal/files/138765735/Brooks_et_al_2017_npj_Science_of_Food.pdf, [accessed August 2018].
3. Horse meat investigation: FSA statement, www.cirmagazine.com/cir/food-agy-statement-horse-meat-investigation.php, [accessed August 2018].
4. Findus, https://en.wikipedia.org/wiki/Findus, [accessed August 2018].
5. Bye, bye Findus: Food firm famed for Crispy Pancakes to be taken off the shelves, https://www.dailystar.co.uk/news/latest-news/491811/Findus-youngs-crispy-pancakes-lasagne-horse-meat-aldi-tesco, [accessed August 2018].
6. Financial sustainability of local authorities 2018, https://www.nao.org.uk/press-release/financial-sustainability-of-local-authorities-2018/, [accessed August 2018].
7. We're entitled to eat safe meat. Why has that become such a lottery?, https://www.theguardian.com/commentisfree/2018/feb/20/meat-food-safety-trading-standards, [accessed August 2018].
8. Food Standards Agency, Monitoring data by year, http://webarchive.nationalarchives.gov.uk/20171207164658/https://www.food.gov.uk/enforcement/monitoring/laems/mondatabyyear, [accessed September 2018].

CHAPTER 3

Is Food Fraud a New Idea?

Given the publicity following the discovery of the horsemeat fraud, you might wonder if food crime is a relatively new phenomenon. It isn't; there is a new desire to label the crimes differently, but food crime of one sort or another has been detected and punished since the 13th century. Was it fraud, the in-vogue term, or adulteration that was punished? In the past, the key term has always been adulteration, because it is easier to prove. Despite several attempts to define the terms, it seems the terms adulteration and fraud have been interchangeable over the centuries.

Since the beginning of time, food was sold or bartered and consequently a 'fair price' had to be agreed to achieve the sale. The purchaser would have wanted a bargain and the seller a profit, so it is reasonable to assume that food may not have always been precisely as labelled to help maximise profit for the seller or, some speculate, to help the cash-strapped consumer afford the food. Whether it was a fraudulent act, poor quality assurance, adulteration or a gesture of goodwill might be debateable. To add clarity to the issue, and in the spirit of fun, I'll examine the examples of food mis-selling outlined in this chapter in a section at the end called Fraud or Adulteration?

There are several references to adulteration of wine and bread which trace the practice back as far as the Greeks and Romans.

Laura Schumm, in her history of food adulteration, suggests that honey, herbs, spices, saltwater, chalk or lead (the latter served as both a sweetener and a preservative) were added to wine.[1] Sugar of lead (lead acetate) was the sweetener of choice for the Romans to add to wine.[2] Pliny (AD70) recorded 'the wheat of Cyprus is swarthy and produces dark bread, for which reason it is generally mixed with white wheat of Alexandria'. I assume that the white wheat of Alexandria was the premium product and therefore attracted a higher price, and thus the bread did too as the consumer at the time preferred lighter coloured bread. He noted, with disapproval, that some bakers kneaded their bread with seawater so that they could save on the cost of salt, and detailed how white earth 'Leucogee' was added to bread to bulk its weight.

To reduce the risk of uprisings resulting from famine, King John declared his first 'Assize of Bread' in 1202 and sought to control the profitability of bakers by fixing the price of bread to the price of wheat. Perversely, this only encouraged the enterprising bakers to increase their profit by increasing the weight of the bread, the theory being the more wheat used the larger, and consequently heavier, the loaf. This adulteration was achieved by the addition of heavier ingredients, namely alum (Leucogee), sawdust or metal, for example. Consequently, the Assize had to be regularly updated, in a game of 'cat and mouse' with the bakers, to include regulations covering the latest method of increasing weight. In 1582, the Assize effectively became the first English Sale of Food Act, and it included details of punishments for first, second, third and fourth offences: a fine, loss of stock, pillory or prison and, finally, banishment from town, respectively. Nevertheless, some bakers never learned. In 1311, Alan de Lyndseye was brought before the Mayor and 'sentenced to the pillory for making bread that was of bad dough within and good dough on the outside'. A short while later, he was back before the Mayor and again sentenced to the pillory for 'selling bread that was made of false, putrid and rotten materials through which those who bought the bread were deceived and might be killed'.[3]

The Middle Ages (11–16th centuries) saw the inception of merchant guilds, powerful groups who controlled the way business was conducted and were responsible for ensuring the quality of products through regulations which they helped draft that applied across Europe. One such guild related to spice merchants. Some spice merchants adulterated their valuable products with juniper berries, seeds, ground nutshells, stones or dust. Following the Reformation (of the Catholic Church in the 16th century), the influence of guilds waned and, consequently, so did their laws.[1]

Prior to the 18th century, there weren't many reliable analytical techniques available to the scientist to enable the identification of adulterants or the assessment of the quality of food. Several people are attributed with leading the fight against poor quality food since the late 1700s – Thomas Wakley, a surgeon and MP, Sir Charles Wood, Chancellor of the Exchequer (1850) and two scientists in particular, Fredrick Carl Accum, a chemist, and Arthur Hill Hassall, a medic, led the way. Accum was the first to raise the alarm on adulteration of food. In 1820, he published 'A treatise on adulterations of food and culinary poisons', and the first edition of 1000 copies sold out within a month. The book was also sold in the United States and translated into German, suggesting the problem was a lot more widespread than London. In the preface, he stated:

'This treatise, as its title expresses, is intended to exhibit easy methods of detecting the fraudulent adulterations of food and other articles.

Every person is aware that bread, beer, wine and other substances employed in the domestic economy are frequently met with in an adulterated state.'

He stated that adulterating materials intended for human consumption with ingredients deleterious to health was the most criminal of offences and had been recorded as the cause of death in some cases.

Accum published the names and addresses of traders convicted for adulterating food and drink; this made him some powerful enemies. He returned to Germany in 1821, but substandard food continued to be sold. Tea and coffee were two popular beverages at the time, and used tea leaves and coffee could be bought for a few pence per pound from London hotels and coffee shops. The used tea leaves were boiled with copperas (ferrous sulfate) and sheep's dung, then coloured with Prussian blue (ferric ferrocyanide), verdigris (basic copper acetate), logwood, tannin or carbon black before being resold. Exhausted coffee was adulterated with other roasted beans, sand/gravel and mixed with chicory, the dried root of wild endive – a plant of the dandelion family. Chicory itself was sometimes adulterated with roasted carrots or turnips. So it seems that even the adulterant was adulterated.[4]

Leucogee or alum $\{K_2SO_4\ Al_2(SO_4)_3 \cdot 24H_2O\}$ continued to be added to bread in the mid-19th century and Dr John Snow associated it with rickets (*The Lancet*, 4 July 1857).[5] His hypothesis was treated with scepticism at the time, the protagonists' argument being that infantile rickets had been endemic across Europe for at least 100 years. We now have a much better understanding of the interactions in the body of aluminium with the calcium and phosphorus metabolisms, so his supposition seems remarkably plausible, particularly as it is clear that the practice of adding alum was in existence for some considerable time before 1857, so the long-standing endemic could have been caused by the adulteration of food using alum. A *Lancet* survey of breads sold in London was carried out in 1851, and alum was found in every sample submitted for analysis.[5]

Between January 1851 and December 1854, Arthur Hill Hassall analysed around 2500 samples in his London laboratory using the previously unused technique of microscopy followed by chemical analysis, and he proved that adulteration was the rule, not the exception (Table 3.1). He also recorded the names and addresses of the vendors and published these details along with the results of his analysis in a book, *A. H. Hassall, Food and its Adulterations; Comprising the Reports of the Analytical Sanitary Commission of 'The Lancet' for the years 1851 to 1854*.[6] The furore of food adulteration continued at a pace with results and issues raised by Hassall being disputed by some in the food industry. A Parliamentary committee was established to assess the accuracy of Hassall's results and many testified, including Thomas Blackwell of Crosse & Blackwell's who gave evidence that the greening of preserved fruits and vegetables with copper salts and colouring of red sauces for potted meats with iron compounds were common. He admitted that his firm used these additives; not realising that they were

Table 3.1 Adulterants found by Hassall and Accum in the 19th century.[4]

Food	Adulterant
Bread	Alum, sawdust, chalk
Red cheese	Coloured with red lead and vermilion (mercury sulfide)
Cayenne pepper	Red lead, vermilion, Venetian red, turmeric, ground rice, mustard seed husks, sawdust, salt, iron, mercury compounds
Pickles and bottled fruits	Coloured green by copper salts
Vinegar	'Sharpened' with sulfuric acid; often contained tin and lead dissolved when boiled in pewter vessels
Confectionery	White comfits often included Cornish clay
	Red sweets were coloured with vermilion and red lead
	Green sweets often contained copper salts (*e.g.* verdigris: basic copper acetate) and Scheele's or emerald green (copper arsenite)
Olive oil	Often contained lead from the presses
Custard powders	Wheat, potato and rice flour, lead chromate, turmeric to enhance the yellow colour
Coffee	Chicory, roasted wheat, rye and potato flour, roasted beans, acorns *etc.*; burnt sugar (black jack) as a darkener
Tea	Used tea leaves, dried leaves of other plants, starch, sand, china clay, French chalk, Plumbago, gum, indigo, Prussian blue for black tea, turmeric, Chinese yellow, copper salts for green tea
Cocoa and chocolate	Arrowroot, wheat, Indian corn, sago, potato, tapioca flour, chicory, Venetian red, red ochre, iron compounds
Porter and stout	Water, brown sugar, *Cocculus indicus*, copperas, salt, capsicum, ginger, wormwood, coriander and caraway seeds, liquorice, honey, Nux vomica, cream of tartar, hartshorn shavings, treacle
Gin	Water, cayenne, cassia, cinnamon, sugar, alum, salt of tartar (potassium tartrate)
Red wine	Juice of bilberries or elderberries

so objectionable.[4] As a result of Hassall's work, the Adulteration of Food and Drink Act 1860 was enacted and subsequently revised in 1872 to incorporate Hassall's proposals, including the naming and shaming of those found guilty of adulteration and the punishment adulterers could expect which, for a second offence, was six months' hard labour.

Adulteration was still not adequately defined, but the 1872 Act declared that the admixture of anything whatever with an article of food, drink, or drug for the purposes of fraudulently increasing its weight or bulk is an adulteration within the previous provision of the Act. The adulteration of intoxicating liquors was covered in the Licensing Act 1872, which provided a schedule of deleterious ingredients which were considered to be adulterations, but sadly this wasn't considered necessary for foods.

Results on approximately 86 000 food samples taken by Local Government Officers in the period 1878–1882 showed that around 15% of foods were adulterated. Key adulterated products were: milk, butter, coffee, mustard and spirits. Proceedings were not taken on all samples because, in many, the adulteration was considered too small. More than 4000 samples of mustard were analysed during this period, with around 10% thought to be adulterated. The adulteration consisted of the addition of flour and turmeric, which was added for the convenience of the consumer who used the product as table mustard, not as a medicinal product (this required the pure form). Pure mustard had an unpleasant bitter taste. Towards the end of the period, Local Government inspectors were aided by undercover officers as the inspectors had become known to traders and, as a result, the inspectors did not always receive the same product which would have been made available to the consumer on the street. In one case, the inspector had been provided with a milk sample to which some cream had been added by the dairyman to make sure of being on the 'safe side'.[7]

Of the 31 605 milk samples taken during the period 1878–1882, 6410 were reported as adulterated. The principal reason was the addition of water. At the time, science did not enable the analyst to pronounce with certainty whether excess water might contribute to infectious disease (due to the unsanitary nature of the water). Results, and indeed prosecutions, were often contested. The tenth Local Government report (1880–1881) stated: *'in some cases the amount of added water was so large as to be, according to the analyst for Plumstead, a serious matter for health, and even the lives of infants'*. In a Salford case where 30% water was found in milk, the farmer said, in his defence, that this was merely owing to the cows having been poorly fed. The analyst remarked that if the defence were true then the matter should be referred to the Society for the Prevention of Cruelty to Animals. Results were disputed in court on a regular basis and there was some uncertainty as to the accuracy of the amounts of added water. There were attempts to set 'Milk Standards' by which the quality of milk could be assessed. The principal chemical officer of the commissioners for the Inland Revenue said (as to Milk Standards).[7]

> '*We have felt ourselves unable to adopt the "definitions" and "limits" for genuineness laid down by the Society of Public Analysts, for simple but all-sufficient reason that they are not borne out by our own analysis of hundreds of samples known to be genuine*'.[7]

This mostly related to the fat content of milk but did cast doubt over other parameters.[7] Consequently, public analysts spent years collating data to support their cases and sharing with others who peer reviewed their results. Birmingham Public Analysts Dr Alfred Hill and J. F. Liverseedge undertook monthly analysis and painstakingly compared the results obtained during the 32 years 1899–1930, analysing 51 703 samples in total and publishing the results.[3,8]

3.1 DID THE PUBLIC KNOWINGLY ACCEPT ADULTERATED FOOD?

Tobias Smollet, a physician and satirist, suggested that the public were aware and accepting that food was of questionable quality when he wrote:

> '*The bread I eat in London is a deleterious paste, mixed up with chalk, alum and bone ashes, insipid to the taste and destructive to the constitution. The good people are not ignorant of this adulteration; but they prefer it to wholesome bread, because it is whiter than the meal of corn [wheat]. Thus they sacrifice their taste and their health to a most absurd gratification of a misjudged eye; and the miller or the baker is obliged to poison them and their families, in order to live by his profession*'.[8]

The advisors to 'Victorian Bakers' (a television documentary),[9] which covered this issue, suggested that the adulteration of bread may simply have been in response to customer demands for cheap food. They argued that, without adulterants, the food would simply not have been available, leading to malnutrition and starvation amongst the poor. At first glance, this does not accord with King John back in 1202, where the public revolted when food was too expensive so he fixed the price; however, this led to adulteration by the bakers to increase their profit. The advisors seem to suggest that adulteration keeps the price down, a similar argument to that presented during Horsegate.

I think the public were aware that food adulteration was prevalent in the mid-19th century; they probably felt there was little they could do to stop it, but they did not want this to be the norm. The well-known cartoon on food adulteration that was published at the time, in the satirical magazine *Punch*, depicted a little girl buying produce for her mother and suggests that she has tea to kill the rats and chocolate to get rid of beetles – thus buying adulterated food for the poisons they contain, suggesting that the public were well aware of the issue.

The public response to Accum's books and lectures (the latter gave him 'celebrity status') would suggest they were interested in protecting themselves, although clearly those who could attend a lecture or buy a book were probably the more affluent and thus had the wherewithal to respond and worry about the issue. In addition, the perpetrators of adulteration who were named and shamed by Accum objected. If they were merely satisfying a demand led by the consumer then this would have been good publicity, why would they complain?

It was suggested during the horsemeat scandal that, by driving down prices, the public and local authorities got what they deserved. I cannot agree with this suggestion; the public are legally entitled to food which is of the nature, substance and quality demanded by the consumer. The law is clear, and there can be no defence that the public really wanted non-compliant food, so long as it was cheap.

3.2 SYNTHETIC FOOD AND ADULTERATION

Some writers, such as Caroline Walker[10] and Bea Wilson,[11] make the argument that foods containing additives are effectively adulterated foods. These foods are enhanced by the addition of substances enabling the manufacturer to improve the look or feel or taste of a product whilst not necessarily using the best quality ingredients, thus increasing profitability. Dr Henry Letheby, 1816–1876, defined adulteration as *'the act of debasing a pure or genuine commodity for pecuniary profit, by adding to it an inferior or spurious article, or by taking from it one or more of its constituents'*. This definition might seem to accord with the views of Walker and Wilson. However, additives such as antioxidants (added to foods that contain fats to stop them going rancid), colours, emulsifiers, stabilisers, gelling agents and thickeners, flavourings, preservatives and sweeteners are permitted and controlled by law and not adulterants. If they were classed as adulterants then synthetic foods such as an instant dessert or, indeed, ready meals, sweets, snacks and numerous other products would be considered adulterated.

Perhaps the question is, if they are not adulterated are they fully compliant with the law? By law (Food Safety Act 1990 and Reg. (EU) 172/2002), food must be safe (not injurious to health), wholesome and not presented in a manner likely to mislead or, to use the phrase from the UK Act, *'of the nature substance and quality so demanded by the consumer'*. Since the additive is permitted, is it presented in a manner likely to mislead the consumer? This I find the most interesting aspect of the debate. How well informed is the consumer? Do they read labels? Research commissioned by the FSA a while ago suggested that the average consumer would buy the groceries for each week in less than 30 minutes and they would buy around 70 items, leaving around 25 seconds to find and select each article. This does not suggest that the average consumer pours over the finer detail contained within the label. Indeed, the finer points of a food label are not that interesting to the average consumer; they place their trust in food businesses and expect the food to be safe, wholesome and to their taste. From a nutritious perspective, this may not be the best strategy, and I believe we need to establish quick systems of communication for the consumer: traffic light labels, for example.

3.3 HAS ADULTERATION CHANGED SINCE THE 19TH CENTURY?

Figure 3.1 shows the relative frequency of Rapid Alert System for Food and Feed (RASFF) notifications for product categories in

Is Food Fraud a New Idea? 53

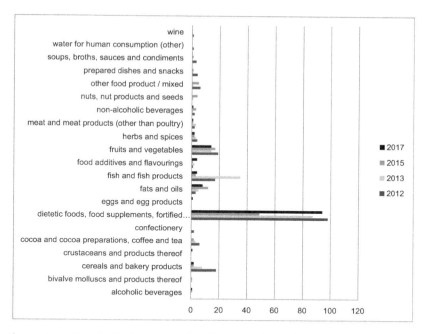

Figure 3.1 Rapid Alert System for Food and Feed (RASFF) notifications reported on risk related to composition by product categories in 2012, 2013, 2015 and 2017.

recent years. These relate to chemical composition only; microbiological issues have not been included to facilitate easier comparison with previous results from the 19th century. It is not possible to compare accurately as techniques and foods have changed considerably over that time; for example, dietetic foods feature prominently now and would not have been available 150 years ago. However, it can be seen that many of the old favourites remain, such as alcoholic beverages, confectionary, herbs and spices, tea and coffee and fats and oils.

3.4 FRAUD OR ADULTERATION?

Let's examine the cases outlined above to ascertain if it was a fraudulent act or adulteration. Please treat this as an interesting intellectual exercise, merely for illustrative purposes. The main caveat is that we have limited information, which in some cases, might be up to 2000 years old. We cannot verify the accuracy of the information, and I am not seeking to besmirch the otherwise good name of the people involved or criticise the courts where the case has been heard in court. If you, the reader, are a distant relative of anyone mentioned or implied in these cases, please don't sue the author as I doubt the royalties from the sales of this book will amount to that much anyway! I am not sure that Sir Bruce Forsyth, the famous Saturday night TV game show host, would have taken this game to Saturday Night TV, but I hope it works for the reader.

3.4.1 Definitions

3.4.1.1 Adulterated Food. A food is adulterated if it is deemed not of the nature, substance or quality demanded by 'a reasonable' consumer, by virtue of the fact it contains an ingredient which should not have been present, either at the concentration found or at any level where the ingredient is not permitted. The adulterant can be a substance (including infestation), whether harmful or not, or an organism which is deleterious to health and the adulterant may arise from the packaging.

3.4.1.2 Fraud. To prosecute fraud, five elements must be satisfied:

- a false statement of a material fact;
- knowledge on the part of the defendant that the statement is untrue;
- intent on the part of the defendant to deceive the alleged victim;
- justifiable reliance by the alleged victim on the statement; and
- injury to the alleged victim as a result.

3.4.1.3 Mitigation
3.4.1.3.1 Due Diligence. The first due diligence case was heard by Stipendiary Magistrate Tim Workman in Westminster in 1997. Since this case many, including me, have grappled with defining due diligence for different scenarios. The truth is this can only be determined in court, and consequently, many of us have an opinion, but until the particular case is heard we

cannot be sure. For illustrative purposes, I would like to use a case prosecuted by my good friend Alan Richards who, at the time, was the public analyst in Durham. The nub of the case involved a complaint that muesli tasted awful (not the actual words used by the complainant). Alan's analysis revealed that the reason why the breakfast cereal tasted the way it did was due to the addition of cat faeces. Further investigations revealed that the shop sold the product direct from open hessian sacks in a 'health food style arrangement'. I am sure you can picture the scene; the shop had previously noted infestation from rodents and used the cat as a 'natural' remedy. It is not hard to predict what might have happened, *i.e.* the cat had used an open sack of breakfast cereal as a litter tray and the complainant, to their disgust, not realising anything untoward, had served this portion in a breakfast bowl and started to consume it. No attempt was made to ascertain how many times the cat had used the open sacks as a litter tray or indeed if it had urinated in the sacks on this occasion, or before. I hope you are not eating as you read this.

Now let's extrapolate: suppose this retailer had a documented system of checking for infestation on a two-hourly basis, with the results being recorded that the sacks were covered when the shop was closed and the cat was not allowed in the shop during opening hours. To date, no infestation had been noted, there were no recorded incidents of complaints of any kind against the company. If you were the magistrate hearing this case, would you consider this diligent? And, as the consumer, would you consider it diligent? Like all cases heard in court, your conclusion will be affected by your perception and level of involvement. Under the circumstances outlined here, I probably would consider the company to have shown sufficient diligence.

3.4.1.3.2 Bought in Good Faith. The 1875 Act (and all subsequent Acts) states that if a retailer can show that he bought an adulterated product in good faith and merely sold it on, then he is not guilty of adulteration or fraud.

3.4.1.3.3 In the Public Interest. In the UK, prior to charging someone with a criminal offence, prosecutors must be satisfied that there is sufficient evidence to provide a realistic prospect of conviction and that prosecuting is in the public interest.

The following example may help to illustrate a public interest test:

- If a 95-year-old lady took an apple from a tree growing over a pavement, is that theft and, assuming she has the necessary intent (to deprive the owner of the apple), would you prosecute for one apple?
- Suppose the evidence suggests that she went into the orchard and picked up the windfall but sold the apples in a market, giving the proceeds to charity; and, finally,
- If she used others to pick fruit and sell for profit.

I suspect that most people would say not in the public interest in the first case, in the second they might be persuaded and, in the third example, they would definitely consider it in the public interest to prosecute (based on the limited information I have supplied).

3.4.2 The Non-compliant Products

We shall take a few examples given earlier in the chapter which, if prosecuted, would have resulted in a charge of adulteration (Table 3.2). For the purpose of this exercise, let us assume that it is in the public interest to prosecute. The possible verdicts might be: fraud, due diligence or adulteration.

It can be seen that the result really does depend on many factors and the burden of proof for fraud is more onerous. For example, proving that a 14th century baker who couldn't read was fully cognisant of the rules of King John's Assize and wilfully deceiving his customers, and therefore committing a fraud, might be more difficult, especially if he sells only a few loaves of bread each day and doesn't keep any records. Thus, adulteration has been the easier option, at least until now.

I hope you found it a thought-provoking and useful illustration. Many readers might be tempted to reach for their keyboards and write to correct errors they spot above, so please remember it was just for fun, no prizes were awarded and hopefully it illustrated the point how difficult it can be to make a decision regarding the right way to prosecute. So much depends

Is Food Fraud a New Idea?

Table 3.2 Examples of cases where the charge of adulteration would have resulted if prosecuted.

Product	Comments	Possible verdict
Wine containing honey, spices and lead	The Romans had a definition of wine which did not allow these additives.	Adulterated
	The manufacturer would know that this was non-complaint and lead was thought to be deleterious to health. NB. Some of the lead content will have migrated from the vessels used. If it could be proven that the retailer was aware.	Fraud
	The retailer might have bought in good faith.	Adulteration for retailer, Fraud for manufacturer
Bread baked during the reign of King John and found to contain heavy ingredients	Given that the Assize had to regularly change to keep pace, then the bakers were probably knowledgeable and selling to the prejudice of the purchaser.	Fraud
Vinegar containing tin and lead dissolved when boiled in pewter vessels	If higher concentrations found and company had undertaken tests and knew yet sold without declaration.	Fraud
	If no evidence of knowledge by manufacturer.	Adulterated
'Sharpened' with sulfuric acid	Sulfuric acid was not a permitted additive and was knowingly added to improve the product and deceive the consumer.	Fraud
Tea Used tea leaves, dried leaves of other plants, starch, sand, china clay, French chalk, Plumbago, gum, indigo, Prussian blue for black tea, turmeric, Chinese yellow, copper salts for green tea	This would be tricky to defend, if it could be demonstrated that the company were buying used tea from hotels and treating it in the manner outlined, with different strategies for different products *e.g.* green tea. They certainly were selling to the prejudice of the consumer and intent could be demonstrated; the only issue might be if there was no legal definition of what tea was at the time.	Fraud

Table 3.2 (Continued)

Product	Comments	Possible verdict
Milk, watered by ~10%	Defence of poor quality assurance practices within the dairy, water incorporated by accident whilst cleaning plant.	Adulteration
Coffee with chicory	Declared Not declared If evidence of knowledge and deliberate selling and misleading.	No crime Adulterated Fraud

on the quality of the evidence gathered during the investigation, the necessary resources for the evidence to be gathered and also taking into account what is in the public interest as well as giving the courts sufficient sentencing powers.

REFERENCES

1. Food Fraud: A Brief History of the Adulteration of Food, https://www.history.com/news/hungry-history/food-fraud-a-brief-history-of-the-adulteration-of-food, [accessed August 2018].
2. List of food contamination incidents, https://en.wikipedia.org/wiki/List_of_food_contamination_incidents, [accessed August 2018].
3. 'Memorials: 1316', in Memorials of London and London Life in the 13th, 14th and 15th Centuries, ed. H. T. Riley (London, 1868), pp. 118–123. British History Online http://www.british-history.ac.uk/no-series/memorials-london-life/pp118-123 [accessed 23 August 2018].
4. The fight against food adulteration, https://eic.rsc.org/section/feature/the-fight-against-food-adulteration/2020253.article, [accessed August 2018].
5. J. Snow, *Int. J. Epidemiol.*, 2003, **32**(1), 336–337.
6. A. H. Hassall, *Food and its Adulterations; Comprising the Reports of the Analytical Sanitary Commission of 'The Lancet' for the years 1851 to 1854*, Longman, Brown, Green and Longmans, London, UK, 1855.

7. T. Herbert, *The Law on Adulteration: Being the Sale of Food and Drugs Acts, 1875 and 1879, with Notes, Cases ... 1884*, Facsimile Publisher, 1884, p. 1.
8. The Expedition of Humphrey Clinker in 1771, Tobias George 1721–1771 Smollett, Andesite Press (8 Aug. 2015), ISBN-13: 978-1297496325.
9. BBC two documentary Victorian Bakers https://www.bbc.co.uk/programmes/b06vn7sj.
10. The Food Scandal, What's Wrong with the British Diet and How to Put it Right, http://www.cwt.org.uk/publication/the-food-scandal-whats-wrong-with-the-british-diet-and-how-to-put-it-right/, [accessed August 2018].
11. B. Wilson, *Swindled: The Dark History of Food Fraud, from Poisoned Candy to Counterfeit Coffee*, Princeton University Press, USA, 2008, vol. 1.

CHAPTER 4

World-wide Food Frauds

The world-wide web of food supply has increased over the past 30 or so years and has resulted in large food supply contracts with suppliers from across the globe. Whilst there are artisan food suppliers who prepare their food locally and sell it to dedicated, often local, customers, there are also multiple outlet food businesses (caterers and retailers) that source their foods from around the globe on a massive scale. It has been estimated that around 50% of UK food is imported. Consequently, when a food crime is committed it need no longer be a local matter, it may be UK-wide and often EU-wide. The consequences can be huge, with far-reaching implications leading to massive product recalls and quick profits for the would-be fraudster. Since the millennium, there have been several food scares which have had a major impact. This chapter will look at some of these issues and discuss whether, in the light of a new desire to prosecute for fraud, a criminal case may have been possible.

4.1 BEEF PIES SUPPLIED FOR UK COUNTY CATERING ESTABLISHMENTS

This case will work as an appetiser; it is relatively small-scale as it affected the county council, who supplied around one million meals per year to people in its care. It may have affected other

The Horse Who Came to Dinner: The First Criminal Case of Food Fraud
By Glenn Taylor
© Glenn Taylor 2019
Published by the Royal Society of Chemistry, www.rsc.org

catering establishments too, although, at the time, the practice wasn't illegal and therefore no crime was committed, other than the fact that we had a contract with suppliers which expressly forbad this practice.

The county council bought ready prepared individual beef pies which required final cooking. The contract expressly forbad the addition of proteins *etc.* from other animal species, which would have been added to boost the apparent meat content. Meat content can be quickly determined by analysing the protein levels in a food. This is far from the best way but, at the most rudimentary level, this test could be used. So, increasing the protein by the addition of, say, a skin from another animal might lead to the analyst thinking that the amount of meat in the pie was higher than it actually was. In other words, the manufacturer in this case could use less beef and add chicken skins to increase profitability.

As part of our routine monitoring, pies were taken from our kitchens (pre- and post-cooking) and tested in our in-house laboratory. The pie was initially tested microbiologically to assess the levels and types of bacteria. In this case, *Salmonella* was found. This caused some concern as this was an unexpected result. *Salmonella* in an uncooked pie might not be a major issue as it would be cooked and thus the bacteria levels significantly reduced. However, in this case, the issue was that this organism isn't normally associated with beef. Consequently, the product was tested for meat content (using a range of tests, not just the protein test mentioned above) and species identification using DNA assessment. The apparent meat content was satisfactory but the results showed that the pie contained both chicken and beef. The supplier was warned and follow-up tests showed the product had met the desired agreed specification. Sadly, the practice returned within a year and the same results were found, and consequently, the contract was terminated. The practice was legal at the time, although the Law has subsequently changed, with two revisions to the relevant regulations since this practice was noted. The latest states the following:

'Added proteins of a different animal origin must now be added to the name of the food for all meat products, not just those that resemble a cut, joint, slice, portion or carcase of meat'.[1]

I do not know if other catering establishments had a similar specification, but I do know that only one other county council monitored its food in a manner similar to us. If others had monitoring and chose only the most rudimentary of tests (protein, for example) then they would have been fooled. This would not have been classed as a crime at the time as this was legal. It might have been classed as adulterated using some of the previous standards (Dr Henry Letheby, 1816–1876, defined adulteration as 'the act of debasing a pure or genuine commodity for pecuniary profit, by adding to it an inferior or spurious article'). Was it fraud? The issue here to determine fraud would be whether this was knowingly and deliberately undertaken. If I was defending, I'd have argued it was non-compliance at the factory as they were able to supply the same product to others and had simply mixed the produce, by accident, on both occasions. If left unchecked, this would have cost the county council as they paid a premium for the product.

4.2 FIPRONIL IN DUTCH EGGS

Fipronil is a nerve agent for insects. It blocks $GABA_A$-gated chloride channels in the central nervous system, and because of its affinity for chloride, it prevents the uptake of these ions, resulting in excess neuronal stimulation followed by death. It is permitted for use in animals not intended for the food chain and used as an insecticide for removing fleas and a pesticide. It is used as a spot application for dogs and cats to remove fleas but no longer for rabbits due to additional toxicity fears. It is classed by the World Health Organization (WHO) as a 'moderately hazardous' class 2 pesticide, with a Maximum Residue Level (MRL) set at 0.005 mg kg^{-1} in eggs and poultry meat,[2] and it is banned for human consumption. It is readily metabolised into fipronil sulfone, which has a similar level of toxicity, and therefore analysis for the major metabolite and fipronil must be undertaken and the results combined and expressed as total fipronil content and compared with the MRL above.

On 2 June 2017, notification was given to the Belgian Federal Agency for the Safety of the Food Chain (AFSCA) by an egg-breaking plant that fipronil was found in eggs supplied to them. On 6 July 2017, the European Anti-fraud platform (AAC-FF) sent an official request for information to the Netherlands. On 20 July 2017, Belgium notified other member states *via* the RASFF system.[3] By August, the Belgian Agriculture Minister Denis Ducarme told a Parliamentary hearing that an internal Dutch document 'reports the observation of the presence of fipronil in Dutch eggs at the end of November 2016.'[4] They argued that the scandal only became known about in August when the Netherlands FSA ordered the withdrawal of the eggs. A great deal of 'finger pointing' has occurred, which is very unusual for the EU authorities. This has mostly been between Germany, Belgium and the Netherlands.

The Belgium notification (on RASFF) suggested that the following areas were affected by this issue: Afghanistan, Angola, Austria, Belgium, Bulgaria, Canada, Cape Verde, Congo (Brazzaville), Cyprus, Czech Republic, Denmark, Equatorial Guinea, Estonia, Finland, France, Germany, Greece, Hong Kong, Hungary, India, Iraq, Ireland, Isle of Man, Israel, Italy, Latvia, Lebanon, Liberia, Liechtenstein, Lithuania, Luxembourg, Maldives, Malta, Montenegro, Netherlands, Norway, Philippines, Poland, Portugal, Qatar, Romania, Russia, Saudi Arabia, Singapore, Sint Maarten, Slovakia, Slovenia, South Africa, Spain, Sweden, Switzerland, Turkey, Ukraine, United Arab Emirates, United Kingdom, and the United States and that levels between 0.0031 and 1.2 mg kg^{-1} were found in eggs.[5] It was suggested by the FSA that around 700 000 potentially contaminated eggs (often not in their shells but used in catering) were exported by the Netherlands to the UK and, as a result, 11 products, including salads and sandwiches, had been withdrawn from supermarket shelves. To date, there have been 111 RASFF alerts relating to this issue. Around six were considered a serious health risk because they contained levels significantly above the safety limit, and two relate to poultry meat. The last alert was notified on 15 March 2018.

Across the EU, fipronil-contaminated eggs were sold in their shells. Around 20 tonnes of contaminated eggs were exported to Denmark alone and millions of eggs were withdrawn across the EU, particularly in Denmark, Germany and Belgium, and 26 of the 28 member states were affected.

The risk was considered to be low. According to the European Food Safety Agency (EFSA)[4] and AFSCA,[6] levels of fipronil below

$0.72~\text{mg}\,\text{kg}^{-1}$ should not cause a safety concern for the consumers. Consequently, any eggs found with concentrations at, or above, this level cannot be sold. The German Federal Institute for Risk Assessment (BfR) calculated the maximum tolerable consumption per day for children (10 kg) and adults (65 kg)[7] and related these levels to how much of a particular food could be consumed. For example, for cookies prepared with eggs containing the reference level above, a child could consume 210 g per day and an adult 1.4 kg per day.[8]

More than 100 Dutch poultry farms were closed during the investigation. In the Netherlands and elsewhere in the EU, poultry farmers employ companies to maintain cleanliness on their farms, and it appears that fipronil was added to disinfectant used on farms by some of these firms despite the ban. The scandal has received much media attention (Figure 4.1) and, as is so often the case, attention moves from the issue rapidly.

A judicial investigation commenced on 8 August in the Netherlands.[9] It has been alleged that this may be fraud[4] and that fipronil was added to another legal disinfectant (DEGA 16), possibly to improve the efficacy.

If there were delays in reporting by either Belgium or the Netherlands, it is understandable as I suspect both wanted to find the source of a fraud, if this was fraud. The key difference in detecting a fraud rather than adulteration is the need to swiftly seize relevant evidence before fraudsters know you are 'on to them'. The key issue for the judicial investigation will be to determine how, and why, fipronil was used, whether this was an error or a deliberate act to sell or use a disinfectant that was more potent without declaring that it contained a banned substance and, indeed, who was knowledgeable about its use. The eggs containing the pesticide above the MRL were clearly not of the nature, substance or quality so demanded by the consumer and failed to comply with EC no 178/2002 and the UK Food (Safety) Act 1990 (although currently one and the same). Were they adulterated? We could examine this using the new definition on the FSA website:

Adulteration – reducing the quality of food by including a foreign substance, in order to lower costs or fake a higher quality.

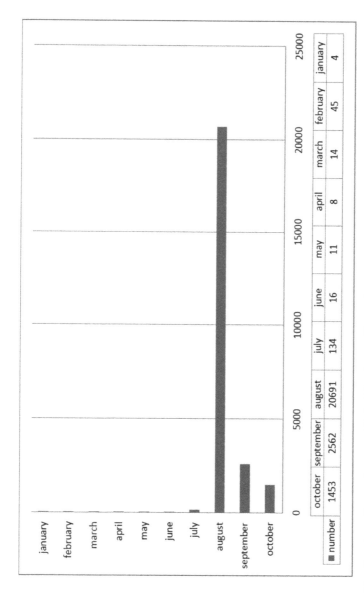

Figure 4.1 Number of news items published by the media on fipronil during January to October 2017. Reproduced from ref. 3.

Fipronil is a foreign substance, and perhaps it was used to lower costs? An interesting conclusion following the judicial inquiry might be that it was simply non-compliant with legislation, but neither adulterated nor fraudulently sold. We shall see.

The latest news in this issue is that EFSA has issued a report (18 April 2018) following the testing of eggs and chicken muscle/fat. EFSA called for an *ad hoc* monitoring programme between 1 Sept 2017 and 31 November 2017, which resulted in 5439 samples. The samples were not only analysed for fipronil but also acarides compounds, which may have been added to remove arachnids (parasites, ticks, mites *etc.*) from chickens. A total of 53 655 tests were undertaken; 742 (13.6%) samples contained residues exceeding the MRL, the legal limit for fipronil.

'Samples that exceeded the legal limit originated from the Netherlands (664 samples), Italy (40 samples), Germany (13 samples), Poland (11 samples), Hungary (6 samples), France (5 samples), Slovenia (2 samples) and Greece (1 sample). Among the 66 substances that were recommended to be analysed in the framework of the *ad hoc* monitoring programme, the only substances found in quantifiable concentrations were fipronil (915 determinations) and amitraz (2 determinations).'[10] EFSA recommend further sampling by member states through their annual programmes.

4.3 GERMAN ASPARAGUS

The Germans are particularly focussed on food miles and the environment, so a locally grown vegetable will be a unique selling proposition and attract a premium price. The claim 'locally grown' is hard to substantiate and easy to claim, especially if paperwork enabling one-up one-down traceability is not available. Frankly, it is just too easy to make this claim and sell 'locally grown' produce on a market stall when generic labels are often used and the produce has travelled much further than the market trader! Of course, it may not be the market trader who makes the claim; it could be a wholesaler, with the trader buying in good faith.

In 2004, enforcement officers in the Munster region could not believe how much locally grown spargel (asparagus) was available. There seemed to be more available on the stalls at any one time than there was space on the local farms to grow it. Clearly something was wrong, but how could you prove it without a method of analysis? The resolution came after discussions between enforcement officers, Dr Axel Preuß (a German food chemist) at CVUA (Chemische und Veterinäruntersuchungsamt) in Münster and Professor Schmidt at the University of Munich. Dr Preuß said, '*We chose isotope ratio analysis focussing on δ^{15}Nitrogen; δ^{13}Carbon δ^{34}Sulfur (from the protein of the asparagus) and δ^{18}Oxygen (from the vapour/head space above the asparagus juice) and determined all four isotope ratios to be on the safe side.*' Stable isotope ratio analysis (SIRA) is complex and expensive, requiring a significant investment. It utilises a principle that elements have stable non-radioactive isotopes which can be identified and quantified to allow calculation of the ratio of the isotope. Comparing this information to the levels of isotopes of carbon, nitrogen and sulfur, it is possible to identify the region from the soil within that region where the food was grown, thus identifying the location of the food. The origin of water used to grow crops can be identified by isotopes of oxygen. Thus, the water trapped in a plant can also be used to identify the region from which that plant had grown.

The UK enforcement authorities would struggle to prove the location of food. I can think of only one public analyst with the resources to detect the location of a vegetable (Eurofins) using this technique.

In 2006, the first year, the authorities found more than 50% of the asparagus samples were not from the home region claimed; they were from France, Spain and other regions in Germany. The fraud rate reduced to less than 5% in 2007 and again in 2008. The traders obviously learned that the authorities can prove the fraud. I am not sure whether the German authorities prosecuted for fraud or non-compliance with regulations. Clearly, the food was mispresented and thus not compliant with the regulations in that it wasn't local and thus a false claim was made; proving fraud would perhaps be more difficult unless paperwork from wholesalers was clear. Unlike the UK, the Germans have prosecuted food fraud for many years.

4.4 CHINESE MILK SCANDAL

Melamine (Scheme 4.1) is an industrial chemical used in plastics. It contains around 67% of nitrogen by weight.

As it is not metabolised, melamine is rapidly eliminated in urine but is not particularly toxic (to mammals or humans), being on par with the toxicity of sodium chloride (salt). However, when it is combined with its analogue cyanuric acid (Scheme 4.2) (also not very toxic but may be a by-product of melamine production), acute renal failure is observed.

4.4.1 Why Add Melamine to Food?

The high nitrogen content of melamine will falsely increase the apparent protein content of a food. Protein will be assessed using the Kjeldahl method, which detects the total nitrogen levels in the food. This is then multiplied by a factor (in the case of milk, 6.38) to ascertain the protein level. The protein level gives an assessment of the apparent milk concentration. 3.1 g of melamine added to 1 L of water can lead to an overestimation of the milk content by around 30%, thus allowing milk to be diluted by around 30%, resulting in a significant profit for the fraudster. It has been suggested that scrap melamine (recycled plastic), which contains cyanuric acid, not the pure substance,

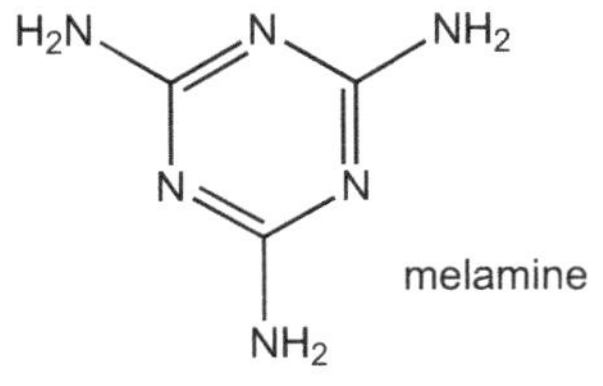

Scheme 4.1

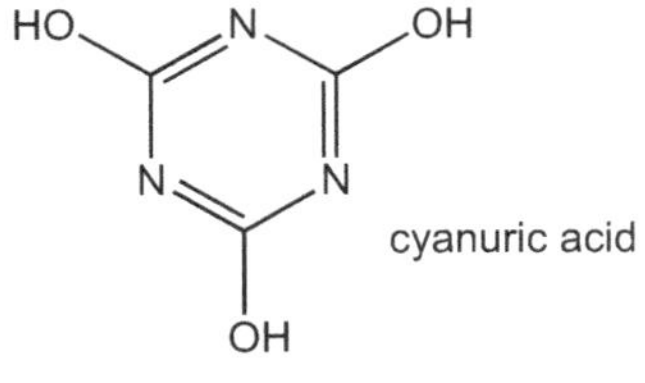

Scheme 4.2

was added to Chinese foods. Farmers had reported salespeople selling protein powders to boost the protein content of milk; this was probably scrap melamine. In China, the price of milk reflects the protein content.

Melamine was first used in baby milk in 2004 in the Chinese provinces of Anhui and Shandong. This was not reported outside of China and is thought to have led to 13 deaths and over 200 babies ill from malnutrition. An analysis of one sample showed that it contained only one-sixth of the required nutrients.[11]

Melamine was first discovered in pet food (Wet Cuts Gravy-style Cat Food) in the United States in 2007, and the EU was informed and all member states were then alerted on the 20 March 2007. It is thought that the pet food had been distributed in the following member states – Belgium, Estonia, Finland, France, Germany, Italy, Latvia, Luxembourg, and Sweden – and Norway and the United States. It was later identified as having been present in wheat gluten (an ingredient in the pet food). Subsequently, 14 further alerts were raised throughout 2007 on the RASFF system. Both wheat and rice protein were found to be contaminated with this compound, causing a large number of dogs and cats to die from renal failure.

It seems to me that the EU had filed this issue in the memory banks. On 10 September 2008, China alerted the world to a problem with baby milk leading to the hospitalisation of infants in the Gansu province. The Chinese authorities reacted at a blisteringly fast pace. Within days, the milk was traced to the Sanlu Group, production ceased, 19 people were arrested, 53 000 babies were recorded as ill and the death toll had risen to four. The head of China's food quality watchdog resigned.

As the news was breaking, I was at the FSA head office and asked how many public analyst laboratories could analyse milk for melamine content. A quick communication with several of my colleagues around England suggested that none of us were set up to undertake this work, let alone accredited (official control labs must be accredited to EN ISO/IEC 17925 for the work they undertake as an enforcement lab). I don't know what the other labs did but, realising this could be a huge problem, Sarah from my lab developed the test and proved its reliability to the accreditation standards within days of my call. We were ready, should there be a need.

The gravity of the situation is all too clear in the statistics below:

- All of the children affected had an exposure to the tainted milk for approximately three to six months before the onset of stones. The highest content of melamine was in Sanlu milk powder, up to 2.563 g kg^{-1} powder, whereas melamine in the other brands ranged from 0.090 to 619 mg kg^{-1}.[12]
- The official data released by the Ministry of Health of the People's Republic of China on 21 September 2008 stated that a total of 52 857 children had received treatment for melamine-tainted milk: 99.2% of the children were younger than three years old.[12]
- 300 000 babies affected, 53 000 hospitalised and six dead.[13]

The first alert from this episode was placed on the RASFF system on 30 September 2007, which was melamine in chocolate cookies reported by the Netherlands; subsequently, 65 further RASFF notifications followed.

Twenty-one people in all were sentenced through the Chinese Courts;[14] these included: Zhang Yujun, who was sentenced to death for producing and selling 776 tons of melamine-laced 'protein powder', and Geng Jinping, also sentenced to death for adding melamine-laced powder to fresh milk and selling to Sanlu and other companies. The following were sentenced to 'life-imprisonment': Tian Wenhua (former chairwoman of Sanlu Group), Zhang Yanzhang (a middleman) and Xue Jianzhong (owner of an industrial chemical shop).[13]

Transition economies, *i.e.* countries that have developed at a rapid pace, do not spend money on infrastructure such as regulation and enforcement at the same rate that their economy grows. The BRICS economies (Brazil, Russia, India, China and South Africa) are classed as transition economies, and they do not have enforcement resources available to monitor compliance, and contracts in such economies can easily become short-term as new alternatives are offered. Others have likened this to the British and US markets of 150 years ago where food adulteration was the rule.

It has been suggested that the government price cap on milk had driven farmers and others towards adulterating the milk in order to survive. In China, the price of milk reflects the protein level, and the price of animal feed and demand for milk had both soared immediately prior to this scandal. Certainly, the sentences handed down by the courts were severe. I do not know if the trials considered fraud or another charge, but I suspect the latter as death sentences were given.

4.5 OLIVE OIL

Olive oil is one of the most frequently adulterated foods. It has been said that this is because it is a liquid which cannot easily be identified by the naked eye, and it is easy to dilute with other oils. It attracts a premium price, especially for 'extra virgin' oil which cannot easily be substantiated. This should be cold-pressed within a few hours of harvesting and have a free-fatty acid content (expressed as oleic acid) of not more than 0.8%. To achieve this acidity, the presses will need to be scrupulously clean.

Olive oil has a unique history and status. It is cherished by many chefs and health conscious people (for its significant health benefits) and has been in use for over 2000 years. It even has a religious status in Christianity and Judaism. The very best oils are given Product of Designated Origin (PDO) or Protected Geographical Indication (PGI) status (like Parma ham) and understandably these attract a premium price. This facilitates another potential fraud, claiming the oil is from a PDO. Extra Virgin olive oil production accounts for less than 10% in many areas and 80% in Greece, 65% in Italy, 50% in Spain, and yet around 80% of oil available in the shops is labelled as 'Extra Virgin',[15] which leaves one wondering!

Around 46% of the world's production of olive oil comes from Spain, and Greece has by far the largest *per capita* consumption, using around 24 L per person per year.[16] It is clearly a big business in Spain and an oil scandal might hurt the economy.

In February 2012, 19 people were arrested: 15 Spanish, two Ecuadorian, one Italian and one Colombian following a year-long joint probe by Spanish Police and Tax Authorities. It was alleged that cheaper oils such as sunflower, palm and avocado were blended in a biodiesel plant and 'adjusted' to hide biomarkers. Police believe that the oils were sold directly in bottles and through unsuspecting third parties.[17]

In November 2017, seven alleged gang members of an organised crime syndicate were arrested in a Greek fake olive oil scandal.[18] Over 17 tons of extra virgin olive oil had allegedly been manufactured from adulterated sunflower oil in a workshop. The oil was sold for around half the normal price using the labels of other producers, which were reported to the Hellenic food authority following complaints in 2015. The Hellenic food authority said producers found their markings on oil that they had neither produced nor sold.

Investigations by Europol show that organised crime is increasingly behind food fraud. On the 3 January 2016, CBS news alleged that organised crime is heavily involved in the adulteration of olive oil. They stated that 50% of olive oil sold in Italy (and around 75% in the United States) was adulterated and that the profit margins were around three times more than that achieved by cocaine dealers; however, the penalties for food fraud are very much smaller.[19] Stuart Shotton, an ex-trading standards officer, warned that the penalties for drug dealing are around 10 years and for food fraud it is around half that, and detection systems are less developed in the food fraud arena.[20]

Even the regulators have struggled. Italian authorities introduced a new rule in 2007 demanding that the farm and the press used to produce the oil had to be declared on each bottle of Italian olive oil. They were trying to stop alleged fraud which included the passing off of Spanish olive oil as Italian olive oil, and other oils such as colza oil (from rapeseed) also passed off as Italian olive oil. The EU took exception to this requirement in 2008 and advised that these requirements cannot be imposed but must be voluntary and that, under EU rules, olive oil containing only a small amount of Italian olive oil can be declared as Italian.[16]

4.5.1 The Spanish Olive Oil Scandal

This story has set the standard for food fraud and olive oil in particular. It is probably the first large food fraud that has been reported and scrutinised by enforcement, scientists and epidemiologists and has been used as 'the food fraud story' for many years. Spanish olive oil was adulterated using industrial oil and this led to a new issue – toxic oil syndrome – but is the story a fake? It seems that olive oil is both a hero and a villain. It has amazing health benefits but is also the food most likely to be subject to fraud, and the regulators struggle to agree with member states how to regulate this prized product, and now we even doubt the biggest fraud story.[21]

On 1 May 1981, an epidemic started which was eventually tracked down to olive oil producers. It killed 1000 and injured 25 000, many of whom were permanently disabled. An eight-year-old boy, Jaime Vaquero Garcia, suddenly fell ill and died in his mother's arms on the way to La Paz children's hospital in Madrid. Learning that his five siblings were also ill, medics called them all into hospital for investigation. Their symptoms were fever, breathing difficulties, vomiting and nausea, followed by pulmonary oedema (the build-up of fluid in the lungs), skin rashes and muscle pain. The family was the first of many to present with the same symptoms; understandably, this made the national news. Medics ascertained that this may be a food poisoning epidemic as the children all lived in tower blocks in the suburbs of Madrid. Medics asked the families to find out what they had all recently eaten and the answer was salads. Mapping the locations of the outbreak, medics thought that the offending produce was probably sold in local street markets (Mercadillos). The market traders moved from area to area, and this allowed the key medic, Muro, to predict the area of the next outbreak. He was correct, the outbreak continued and in the area predicted by Muro. He travelled the markets and found unlabelled containers of cooking oil. Some samples were taken, along with samples from the homes of the affected families. These were sent for analysis. On 10 June, an announcement from the government confirmed that they had found the culprit: contaminated cooking oil. Muro was amazed as the results of the tests on the different oils showed no consistency, just a mixture of different oils, including olive oil. This could surely not be the cause? There would need to be consistency in the analysis if the oil was the cause. To protect its olive oil industry, the government had previously banned imports of rapeseed oil (known locally as colza). Streetwise entrepreneurs had imported the oil anyway, selling it through markets *etc.*, some removing the aniline. The government attributed the illness to aniline poisoning.

Three weeks later, the government health teams asked for samples of oil from families in the area who were affected to be brought into the ministry in exchange for free oil from the government. These samples would be analysed to ascertain the true cause. Some suggest that accurate records were not kept regarding the source of these samples and that some people who weren't affected submitted any oil they could find just to receive the free olive oil in exchange. Unsurprisingly, the analysis gave a wide range of results, including those which suggested that some families may have consumed industrial oils.

In 1983, at a Madrid conference, the phrase toxic oil syndrome was first used as the cause of the epidemic. In 1985, Sir Richard Doll (the UK epidemiologist) said that if one family had not had contaminated oil then the cause could not be the oil, there needed to be consistency. He gave evidence in a court hearing a while later where he stated that, in the light of fresh epidemiological evidence, the oil was the source of the outbreak. Judges stated in their summation that they couldn't be certain what the toxin was in the oil and they gave lengthy jail sentences to the oil merchants.

Scientists continued to examine the evidence, and it would appear that the same oil had been sold elsewhere – throughout Spain and in France – with no adverse effects. Many of the families affected said they hadn't bought the oil, and one patient contracted the disease at least 18 months before the start of La Colza. These facts all throw doubt on the cause of the outbreak. So, if it wasn't the oil, what was the cause? The suggestion from Muro and others is that the true cause of the outbreak was organophosphate pesticides sprayed on tomatoes which were consumed in the salads. Certainly, the symptoms would match that suggestion.

It seems there will always be two schools of thought on the reason for the outbreak, but one final piece of evidence is the fact that, unbeknown to the authorities, the families affected independently took samples of their own oil to be analysed by one of Spain's most prestigious labs. The results are still available and showed that the oil was not rapeseed, but olive oil.

We (students of food fraud) all still refer to the toxic oil syndrome caused by the Spanish oil fraud. It seems that there is more than a little doubt about the true cause of this outbreak, and even the olive oil fraud story may be a fraud.

However, in conclusion, if all you have to do is add a little Italian olive oil to a bottle of cheaper inferior oil, it's hardly surprising that this is the most fraudulent of foods. If anything, we invite fraud if we do not set the right regulations. The mystique around olive oil continues.

4.6 FRENCH WINE FRAUD

Another valuable liquid that attracts a price premium is wine, and the top French wines attract some of the biggest premiums. They will therefore always be on the radar of the fraudster, be it the opportunist wine waiter who simply swaps the bottle or label before serving, the organised crime syndicate or someone inside the organisation.

A breaking story (March 2018)[22] suggests that the insider in one of the biggest wine frauds might have been right at the top of a bottling company that passed off 66.5 million bottles of cheaper wine as three of the biggest, best known brands of French wine: Côtes du Rhone, Côtes du Rhone Villages and even Châteauneuf du Pape. Apparently, the wine is worth around £4 per bottle until it is relabelled, and then it apparently sells for around £50.[23] According to Virginie Beaumeunier, the CEO of France's consumer protection body, Direction Générale de la Concurrence, de la Consommation et de la Répression des Fraudes (DGCDRF), the *'CEO of the company'* was *'indicted for deception and fraud'*.[22] Could the thief who came to dinner be the host in this case? The DGCDRF have not confirmed the name of the company or the CEO concerned but many in the media have. It is alleged that the fraud has been running since October 2013 to March 2017 and that the company sourced four million cases of table wine and sold it as premium wine.

In an unconnected case in 2016, French wine baron Francois-Marie Marret was jailed for two years and fined £7 million for passing of poor quality wine as Saint-Emilions, Pomerols and Listrac-Medocs (his own labels).[24] The so called 'moon wines', named because they were delivered to the Barron by night, were blended with his own high end wines and sold to his clients: some of the biggest retailers in the EU netting around 800 000 euros for:

- the baron;
- a wine merchant, Vincent Lataste (18-month suspended sentence and £4500 fine; his company was fined £27 000, half of which was suspended);
- an employee of QualiBordeaux, an independent quality control body (one-year suspended sentence);
- a broker (eight-month suspended sentence);
- growers (six-month suspended sentences and up to £11 000 in fines); and,
- a driver who delivered the wines from one chateau to another by night (four-month suspended sentence).[24]

In New York, a very well-known wine merchant held his April 2008 sale at a New York restaurant aptly named Cru.[25] Early on in the auction, two bottles of Dom Pérignon Rosé sold for $42 350 each, setting a new record. It seemed it would be a good evening for both sellers and buyers. Rudy Kurniawan had a consignment of 268 bottles of Domaine Armand Rousseau, Domaine Georges Roumier and Domaine Ponsot wine for sale. He was well known as a specialist dealer in the finest burgundies that the Domaine de la Romanée-Conti region had to offer. At one stage, it was rumoured that he was spending $1 million per month on wines like these. He had previously sold consignments through the same auctioneer, Acker, Merral & Condit, on two previous occasions, netting around $35 million.[26] Just as the auction started, Laurent Ponsot, the proprietor of Domaine Ponsot, arrived to stop it, or in particular, the sale of 97 bottles allegedly bearing his name. He had expressed concerns that some of the bottles were counterfeit, in particular the 1929 bottle of Clos de la Roche, which the company didn't start producing until 1934, and 38 bottles of Ponsot Grand Cru, Clos Saint-Denis which started production in the 1980s, although the bottles presented at auction were labelled as from 1945–1971. The Ponsot bottles were withdrawn from sale at the beginning of the auction. Laurent Ponsot was understandably less than happy. His family name was at risk of becoming a toxic brand, and he decided to get to the bottom of the issue and clear the family business. This took four years and involved the FBI. On 8 March 2017, Rudy Kurniawan, 41, was arrested and charged with fraud in connection with the sale of bogus Ponsots.[27] When agents searched Kurniawan's house, they found a counterfeiting factory with fake bottles and labels from the most prestigious wines of Burgundy and Bordeaux and cheap Nappa wines. It seems he may have had millions of dollars of counterfeit wines, ripping off some of the most affluent collectors and damaging the reputation of vintage wine forever, but not the name of Ponsot. The auction house is still owed millions, which it lent to Kurniawan against bottles it held on his behalf, and according to the documentary film 'Sour Grapes' about his fraudulent lifestyle, there may be as many as 10 000 bottles of his fake wines still in private collections.

In 2010, 12 winemakers and dealers in Languedoc-Roussillon were convicted of selling millions of bottles of fake pinot noir to E&J Gallo, the US wine brand. Their defenders noted that no American customers had complained.[23] Perhaps this only illustrates how difficult it is for the consumer to be aware of food fraud and how easy it is to defraud. It certainly seems that, in the case of wine, fraudsters will be invited to dinner to win the trust of their targets; they may even be the hosts.

4.6.1 Breaking News

A story about 'Replica Wines' was reported in *The Times* newspaper on 21 July 2018. This issue has been coming for some time; many times during my career I was asked by potential clients to analyse a successful branded product and supply a complete breakdown of the major constituents. I guess the clients wanted to mimic the product, taking some of the loyal customers from that brand and sell theirs as the cheaper alternative which is as effective. I never accepted any of those offers. How could I? We worked for the prosecution not for industrial intelligence purposes. Now a US company has been formed, Replica Wines, the brain-child of a wine merchant and a food scientist. Food technologists analysed a famous wine and identified the chemical components responsible for the flavour, aroma and colour *etc.*; 580 chemicals in total. They then blend a base wine (a bulk wine from the region), adding acidity, flavours, wood essences, and adjust the alcohol and sweetness to match the target wine. Once made in the laboratory, it is indistinguishable from the target wine, which retails at approximately twice the price of the copy. The replica is sold under names such as 'Knockoff' or 'Pickpocket'; the label does not mention the target wine, although the accompanying sales information might. 'Replica Wines' argue that even the best wines are blended by expert vintners to achieve a consistent product in keeping with their customers' expectations and that this process is hardly different from theirs in that sense. Is it fraud? No, not if the wine is clearly labelled and the customer is aware of what they are buying; then there has been no deception, therefore no fraud. Is it a rip off, and where does this differ from other products such as strawberry blancmange that has never contained a strawberry, just the flavour? I am sure we will hear more regarding copyright, and no doubt the vintners who have spent many years developing their brand will be aggrieved and challenge through the courts, but this is not significantly different from other examples where products of designated origin like 'Parmigiano Reggiano' cheese are mimicked and sold as parmesan, for example.

4.7 FISH FRAUD IN THE UNITED STATES AND WORLD-WIDE

You might think fish would not be a target for the fraudster. After all, its morphological traits, such as shape and skin colours *etc.*, are a perfect way of identifying the species; at least they are when the fish is presented whole on a fishmonger's slab. If you thought this, perhaps think again.

The EU has told some member states that the practice of bleaching fish must stop as it contravenes EU regulations. Hydrogen peroxide, the solution used to bleach fish, may not pose a health risk but the EU says that the widespread practice of bleaching dark fish meat such as octopus, squid and cuttlefish to make them more attractive on market stalls and to appear fresher than they are contravenes EU law. The EU is concerned that there is room for scams, particularly as the consumer cannot tell the difference between a treated and non-treated fish. The EU Commission states that *'hydrogen peroxide does not appear on the list of food additives authorized in foodstuffs contained in Annex II of regulation no. 1333/2008. The use of this substance as a food additive is therefore not authorized in the European Union. The Community executive notes that the responsibility for the member states to effectively enforce Union legislation relating to the food supply chain, which also includes rules applicable to the use of food additives, is incumbent'.*[27] Clearly, those undertaking this process were not declaring it at the point of sale.

Now, if your white fish is nearing its sell-by date, fear not. It can be rejuvenated by soaking it in a solution of citric acid and hydrogen peroxide until the fish bleaches in colour and returns to the glory of its previous appearance, *i.e.* much whiter and fresher in appearance. The hydrogen peroxide might not be a health issue, but the levels of histamine and nitrates now present as part of the decay process are still high in the rejuvenated fish and can affect those with respiratory difficulties, for example COPD and cardiovascular disease, and is one of the most common toxicities related to fish, constituting around 40% of all seafood-related illnesses reported to the US Centers for Disease Control (CDC).[28] At a meeting to share expertise on food fraud in Rome in October 2014, I heard an excellent presentation (case study no. 3) from General Cosimo Piccinno, the Commander of NAS (the Italian Food Police). He advised the audience that the NAS had just busted a company bleaching old fish in the manner outlined above. Their operation was to disguise the ageing of the fish and sell it at markets using an extended shelf-life. He expressed concern that those with cardiovascular problems may suffer as a result of eating this fish. He alluded to the fact that organised crime might be behind operations such as this. His presentation was delivered with all of the aplomb one would expect of a well-respected long-serving member of the Italian Carabinieri. Sadly, in July the following year, the General died following a battle with cancer. He will be greatly missed, having given his life to enforcement and the protection of health.

If fraudsters are not bleaching fish, then they colour it using dyes. It is alleged that, each week, five million portions of tuna that would be destined for the tin due to their colour are adulterated and sold as fresh for twice the price, costing consumers £174 million.[29] Vegetable extracts containing high levels of nitrates are illegally used to change the colour of the tuna. As a result, the consumer eats raw tuna that is not fresh and contains elevated levels of nitrates and histamines, which can lead to the health problems noted above. In 2013, an Oceana survey of all tuna species sold by restaurants and retailers all around the United States revealed that, in 71% of the samples analysed, lower grade fish had been substituted for the more expensive tuna the customer thought they were buying. It seems that sushi restaurants may lead the way in substitution of lower quality tuna for the more expensive variety on the menu, and they also offer white tuna, which doesn't exist as a species and may be escolar, a species of fish which is banned in Japan and Italy.[29]

On the other side of the world, pargo (*Lutjanus purpureus* or Caribbean red snapper) is the most economically important snapper in Brazil. During the process of filleting, there is an opportunity to defraud and, yes, you have guessed, cheaper red snapper can be used in addition or instead. In Brazil, two species can be labelled pargo, which attracts the premium price. A recent survey carried out between March 2013 and Octoberr 2014, with samples taken from supermarkets in the north of Brazil which were processed by a single supplier, showed that 22% of fish samples analysed by DNA techniques were not pargo but *Rhomboplites aurorubens*, a snapper with a commercially lower value.[30] This is far from unusual as, once a fish is filleted, it seems it becomes fair game for fraudulent practices, particularly it seems if it is sold to catering establishments. Seven years ago, Oceana launched a new campaign in the United States called 'Stop Seafood Fraud'. Oceana are the largest advocacy group working to protect the world's oceans. In a survey at the time, they analysed well over 1000 fish from 50 different cities in the United States using DNA identification techniques to show that around 50% of the filleted samples they analysed were not the species on the menu in the restaurant.

Oceana's 2015 survey says, in 43% of samples taken, salmon was mislabelled as line caught when it was actually the cheaper product, farmed salmon.[31] Most of this fraud (79%) happens in the restaurant rather than the retailer. Figures 4.2 and 4.3 show the fraud uncovered by Oceana[32] and reported in the *Huffington Post*.

World-wide Food Frauds

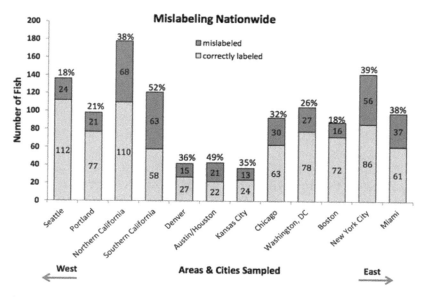

Figure 4.2 Overall mislabelling rates for various metropolitan areas or regions across the United States in 2013.
Reproduced from ref. 32 with permission.

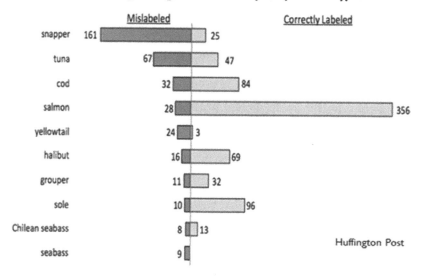

Figure 4.3 Mislabelling rates among most commonly sampled fish types. Mislabelling rates vary greatly between the type of fish purchased.
Reproduced from ref. 32 with permission.

The United States consumes more fish than almost everyone else (excluding China). 90% is imported and less than 1% tested for fraud using DNA or similar methods. It is simply too easy to defraud the catering trade if the fish isn't going to be tested. There is more chance of the larger retailers undertaking testing, and likewise a chance that enforcement officers might check the fish sold on the street, but many in the catering trade and suppliers seem to be happy to take a chance. This needs to change if fraud is going to stop.

A recent report by Alan Reilly for the United Nations Food and Agriculture Organization states that fish is vulnerable to fraud and has been for many years. The practice of fraud in this area is widespread and a great deal more effort is needed. It is possible to trace timber from all over the world using barcode tracer systems, and it should be possible to use digital tracer systems for fish. This would impact on the ease of operation for fraudsters in this arena.[33]

4.8 CHILLI POWDER FROM INDIA

Probably the most well-known food scare in recent years happened in 2005 and concerned chilli powder from India coloured red using an illegal dye called Sudan 1 Red, an azo-dye which is thought to be associated with cancer. This caused the biggest recall in history and amounted to around 600 products. Italy raised the alarm, finding the dye in Worcester sauce. The surprise for many was how many foods contained Worcester sauce. By the time the contaminated sauce was used, the levels found were very small, but the dye is not permitted so, whilst there was very little risk of harm, a precautionary approach was taken by the FSA. The dye is normally used in wax, petrol and floor polish and was thought to have been added to chilli powder in India before it was exported.[34]

REFERENCES

1. The Meat Products Regulations 2014 Guide to Compliance, 23rd January 2014, https://consult.defra.gov.uk/food/meat-products-england-regulations-2014/supporting_documents/MPR%20technical%20guidance.pdf, [accessed August 2018].

2. EU Regulation (EC) 396/2005 of the European Parliament and of the Council of 23 February 2005 on maximum residue levels (MRL) of pesticides in or on food and feed of plant and animal origin for fipronil, http://data.europa.eu/eli/reg/2005/396/oj, [accessed August 2018].
3. Fipronil In Eggs European Commmission Factsheet, http://publications.jrc.ec.europa.eu/repository/bitstream/JRC110632/jrc110632_final.pdf, [accessed August 2018].
4. Belgium accuses Netherlands of tainted eggs cover-up, https://luxtimes.lu/archives/1360-belgium-accuses-netherlands-of-tainted-eggs-cover-up, [accessed August 2018].
5. RASFF Search Portal, https://webgate.ec.europa.eu/rasff-window/portal [accessed September 2018].
6. Conclusion on the peer review of fipronil, *EFSA Scientific Report* (2006), 65, 1–110 (https://efsa.onlinelibrary.wiley.com/doi/pdf/10.2903/j.efsa.2006.65r).
7. Risk assessment and risk management with regard to the presence of fipronil in eggs, egg products, poultry meat and processed products; http://www.favv-afsca.be/businesssectors/foodstuffs/incidents/fipronil/_documents/NoteFipronil_17.08.17_ENG_v1.1.pdf, [accessed August 2018].
8. Fipronil in foods containing eggs: Estimations of maximum tolerable daily consumption, http://www.bfr.bund.de/cm/349/fipronil-in-foods-containing-eggs-estimations-of-maximum-tolerable-daily-consumption.pdf [accessed September 2018].
9. 2017 Fipronil eggs contamination, https://en.wikipedia.org/wiki/2017_Fipronil_eggs_contamination, [accessed August 2018].
10. European Food Safety Authority (EFSA), H. Reich and G. A. Triacchini, *EFSA J.*, 2018, **16**(5), e05164.
11. China 'fake milk' scandal deepens, http://news.bbc.co.uk/1/hi/world/asia-pacific/3648583.stm, [accessed August 2018].
12. A. K.-C. Hau, T. H. Kwan and P. K.-T. Li, *J. Am. Soc. Nephrol.*, 2009, **20**(2), 245–250.
13. 2008 Chinese milk scandal, https://en.wikipedia.org/wiki/2008_Chinese_milk_scandal, [accessed August 2018].
14. Timeline: China milk scandal, http://news.bbc.co.uk/1/hi/7720404.stm, [accessed August 2018].
15. Can you tell if your olive oil is REALLY extra virgin? MailOnline visits Spain and discovered 80% of the

industry if fraudulent, http://www.dailymail.co.uk/travel/travel_news/article-3699232/Can-tell-olive-oil-REALLY-extra-virgin-MailOnline-visits-Spain-discovers-80-industry-fraudulent.html, [accessed August 2018].
16. Olive Oil, https://en.wikipedia.org/wiki/Olive_oil, [accessed August 2018].
17. Olive Oil Times, Europe's Origin Labels, https://www.oliveoiltimes.com/olive-oil-basics/olive-oil-varieties/pdo-pgi-tsg/2238, [accessed August 2018].
18. Olive Oil Times, Seven Held in Greece for Alleged Olive Oil Fraud, https://www.oliveoiltimes.com/olive-oil-business/seven-held-greece-alleged-olive-oil-fraud/60015, [accessed August 2018].
19. Whitaker Bill. (3 January 2016). *"Agromafia"*. 60 Minutes. Retrieved 3 January 2016. 10-minute video & *"Mafia Control of Olive Oil the Topic of '60 Minutes' Report". Olive Oil Times. 3 January 2016.* Retrieved 28 January 2016. Summary of CBS video.
20. Criminals drop drugs for food fraud, https://www.foodmanufacture.co.uk/Article/2011/11/23/Criminals-drop-drugs-for-food-fraud, [accessed August 2018].
21. Cover-up, https://www.theguardian.com/education/2001/aug/25/research.highereducation, [accessed August 2018].
22. Massive Côte du Rhône fine-wine fraud uncovered by French police, https://www.thelocal.fr/20180316/massive-cte-du-rhne-fine-wine-fraud-smashed-by-french-police, [accessed August 2018].
23. Vintage scam: 66 million bottles of French wine said to be fake, https://www.thetimes.co.uk/article/vintage-scam-66-million-bottles-of-french-wine-said-to-be-fake-nwrswg0vp, [accessed August 2018].
24. Wine boss jailed for two years and fined £7m after selling inferior Bordeaux as upmarket labels to supermarket chains, http://www.dailymail.co.uk/news/article-3903852/Wine-boss-jailed-two-years-fined-7m-selling-inferior-Bordeaux-upmarket-labels-supermarket-chains.html, [accessed August 2018].
25. A Vintage Crime, https://www.vanityfair.com/culture/2012/07/wine-fraud-rudy-kurniawan-vintage-burgundies, [accessed August 2018].

26. Rudy Kurniawan, https://en.wikipedia.org/wiki/Rudy_Kurniawan [accessed August 2018].
27. Bruxelles avverte Roma: è vietato sbiancare i calamari con l'acqua ossigenata, http://www.lastampa.it/2016/06/15/esteri/bruxelles-avverte-roma-vietato-sbiancare-i-calamari-con-lacqua-ossigenata-eYHagX33Tgj4QZ17UO9b1J/pagina.html [accessed August 2018].
28. Histamine Toxicity from Fish, https://emedicine.medscape.com/article/1009464-overview [accessed August 2018].
29. Fraudsters are dyeing cheap tuna pink and selling it on as fresh fish in £174M industry, https://www.independent.co.uk/life-style/food-and-drink/dyeing-cheap-tuna-pink-sell-fresh-fish-fraud-174-million-industry-any-other-key-words-a7532066.html [accessed August 2018].
30. I. Veneza, R. Silva, L. Freitas, A. Silva, K. Martins, I. Sampaio, H. Schneider and G. Gomes, *Neotrop. Ichthyol.*, 2018, **16**(1), e170068.
31. How Seafood Fraud Tricks Consumers Into Buying Lower Quality Salmon, https://www.huffingtonpost.co.uk/entry/salmon-fraud-is-rampant_us_597f435de4b02a8434b7ebdc [accessed August 2018].
32. Oceana Study Reveals Seafood Fraud Nationwide, http://oceana.org/sites/default/files/reports/National_Seafood_Fraud_Testing_Results_FINAL.pdf [accessed August 2018].
33. Overview of Food Fraud in the Fisheries Sector, http://www.fao.org/3/i8791en/I8791EN.pdf [accessed August 2018].
34. Food dye scare sparks largest recall in history, https://www.telegraph.co.uk/news/uknews/1484071/Food-dye-scare-sparks-largest-recall-in-history.html [accessed August 2018].

CHAPTER 5

Off With Their Heads

'Off with their heads' cried the Queen of Hearts in another of her furious rants. 'Does she say that frequently?' asked Alice. The White Rabbit replied, 'All the time.' I hope Lewis Carroll will not mind me borrowing, albeit inaccurately, from *Alice's Adventures in Wonderland,* but it helps illustrate a point. Until now, the UK public had been fairly ambivalent about all food scares. No media scrum for heads to roll and no politician shouting from the rooftops. There were four key factors that made this different from other food scares:

1. They made us eat our pets, how could they?
2. Politicians did call for heads to roll, politicians at the highest level: 'Horsemeat scandal: David Cameron says offenders will feel full force of law'.[1]
3. Chris Elliott would not allow his report to be published on a Friday afternoon before recess of Parliament only to be forgotten after the holidays. 'Take all eight pillars or take none', he might have cried.
4. They backed the creation of a new food crime unit and all of the eight suggestions. Elisabeth Truss, the Environment Secretary, said that all suggestions in Chris Elliott's report would be accepted.[2] This included the new food crime unit, a specialist unit which 'upped the ante' on food crime, a

---

The Horse Who Came to Dinner: The First Criminal Case of Food Fraud
By Glenn Taylor
© Glenn Taylor 2019
Published by the Royal Society of Chemistry, www.rsc.org

hitherto unknown concept, and would enable enforcement to prosecute for the crime of fraud. Now it needed clarity and sentencing guidelines for judges[3] and clarity for those involved in prosecuting.

5.1 THE NEW GUIDELINES

Until now, there has been limited guidance for judges and magistrates when dealing with complex and serious offences that do not come before the courts as frequently as some other criminal offences. As a result, there were few 'yardsticks' for the courts to use to help determine a sentence. Now offenders may receive higher penalties, particularly large organisations committing serious offences. It is not expected that higher fines will be levied in all cases.

Rod Ainsworth, Director of Regulatory and Legal Strategy at the FSA, said:

'We welcome these guidelines. They will ensure that there is consistency in sentencing for food safety and food hygiene offences across the country. They will also ensure that offenders are sentenced fairly and proportionately in the interests of consumers'.

The guidelines cover the following offences (*inter alia*):

- Food Safety and Hygiene (England) Regulations 2013, regulation 19(1).
- Food Hygiene (Wales) Regulations 2006, regulation 17(1).
- The General Food Regulations 2004, regulation 4.

5.2 TIMETABLE OF EVENTS

January	2013	Horsemeat fraud identified.
July	2014	Chris Elliott report calls for clear laws.
March	2015	Boddy and Moss sentencing hearing – forgery and traceability breaches.
May	2015	Farmbox sentencing hearing – not horsemeat but inadequate record keeping.
November	2015	New sentencing guidelines published.
February	2016	New guidelines for sentencing in force.
July	2017	Horsemeat trials using the new guidelines.

Caveat emptor (buyer beware if you bought the book) or reader beware. This is intended as a simple brief overview of law relating to food and is not intended as legal guidance.

In the EU, Regulation (EC) No 178/2002[4] lays down the general principles and requirements of food law and the procedures for matters of food safety. In a nutshell, the key parts of interest which food businesses must adhere to are as follows. They must:

- Place safe food on-the-market, *i.e.* food must not be injurious to health or unfit for human consumption. A food may be rendered injurious to health by:
  ○ adding an article or substance to the food;
  ○ using an article or substance as an ingredient in the preparation of the food;
  ○ abstracting any constituent from the food;
  ○ subjecting the food to any process or treatment;
  ○ with the intention that it shall be sold for human consumption.
- Maintain records relating to the traceability of food, *i.e.* identify their suppliers of food, food-producing animals and any other substance intended, or expected, to be incorporated into food; identify the businesses to which they have supplied products; and produce this information to the competent authorities on demand.
- Label, advertise and present food so as to not mislead consumers:
  ○ the food must not be falsely described;
  ○ the food must be of the nature or substance or quality demanded by the purchaser. This allows for the prosecution of adulterated food.
- Unsafe food must be withdrawn from sale or recalled from consumers if it has already been sold. This includes notification of competent authorities where unsafe food is already placed on-the-market.
- To ensure food imported into, and exported from, the EU shall comply with food law.

In England, The Food Safety and Hygiene (England) Regulations 2013 and The General Food Regulations 2004 provide additional local detail on the interpretation of the EU law and update the Food Safety Act of 1990. Food hygiene legislation is closely related to the legislation on the general requirements and principles of food law but specifically relates to the microbiological quality of food. The legislation lays down the food hygiene rules for all food businesses, applying effective and proportionate controls throughout the food chain, from primary production to sale or supply to the food consumer. The defence of 'Due Diligence' includes concepts of 'effective systems or controls' and 'proportionality'. In addition, under section 21, Part 2 of the Food Safety Act 1990 (UK), it is a defence for a food business operator to prove that he took all reasonable precautions and exercised due diligence to avoid the commission of the offence, *i.e.* due diligence defence.

5.3 WORLD-WIDE HARMONY OF FOOD LAW

The World Trade Organization (WTO) seeks to ensure fair trade amongst its members and operates in most areas of the world where free trade is championed, *i.e.* nearly every country across the globe. At the heart of consumer protection law (a group of regulations) is fair trade and protection of consumers. These regulations are designed to stop one firm from gaining a competitive advantage over another by entering into illegal acts such as fraud, and deception of non-compliant products, and to protect the consumer, including the provision of additional requirements to protect the most vulnerable. The Consumer Protection Act covers products including food. Consumer protection law is based on Strict Liability, *i.e. Actus Reus* or 'guilty act'. A person or organisation is responsible for the consequences irrespective of criminal intent or fault, *i.e.* it is an offence to place a product for sale which does not comply with legislation. If the commission of an offence is due to the act or default of another person within the organisation – for example, an employee – then the organisation is guilty of the offence. Thus, there is a harmony of approach towards consumer protection across the globe, at least in terms of strategy; however, local differences apply, which adds to the confusion. The key problem is consistency. For example:

- In the EU, irradiation of food to preserve the shelf-life and ensure the microbiological quality of food is permitted, but

it must be declared. In the United States, it need not be declared on the label.
- Around the world, antibiotics are permitted for the treatment of animals. Food presented for sale, however, must not contain any antibiotics. Therefore, milk from treated animals, or the animals themselves, must be withdrawn from the food chain. For example at an EU border check, crustaceans from India were withdrawn when they were found to contain chloramphenicol, an antibiotic used to protect farmed prawns from infections.
- Attitudes towards genetic modification in foods in the United States are very different from that in the EU. In the EU (including in the UK), foods must say on their label if they contain or consist of genetically modified organisms (GMOs) or contain ingredients produced from GMOs. Very few foods containing GMOs are available for sale in the UK due to attitudes towards these 'Frankenstein Foods', even though this is not based on sound scientific principles. In the United States, they concluded that foods containing GMOs are not fundamentally different from conventional foods in terms of overall composition. Therefore, they do not need legislation specifically for dealing with GM foods. These foods fall under the Food and Drug Administration (FDA) classification of 'generally recognised as safe', and thus they do not normally require labelling or need to be approved before entering the market.
- One of the key differences between US and EU food law is the fact that, in the United States, manufacturers are expected to self-determine if their products are safe and therefore can be granted GRAS status (generally recognised as safe). In the EU, this is awarded centrally by the EFSA.
- One item which reached newspapers recently was the washing of chicken to remove pathogens using chlorinated water.[5] This practice is accepted in the United States but banned in the EU. Currently, the EFSA is considering the use of peroxyacetic acid instead. Given the recent concerns of *Campylobacter* on chicken carcasses, perhaps this should be carefully considered. The EU principle of zero tolerance does not always serve consumers as well as it might.

5.4 THE FDA FOOD SAFETY MODERNIZATION ACT (FSMA) UNITED STATES

The United States introduced the Food Safety Modernization Act (FSMA) under President Obama in January 2011. It focusses on stopping food safety issues (foodborne illness) before they occur, still supporting the ethos of self-determination of safety but giving tools for manufacturers to use and seven major rules for them to follow throughout their supply chain, including risk assessments for importers to assess that the foods they import have been produced in a manner that meets applicable US safety standards. The FSMA requires all food businesses to register with the FDA, and a similar requirement is in place throughout the EU. In the United States, however, the businesses pay for registration and continue to pay an annual fee of $500 to help fund the costs of enforcement. Consultants will help food businesses prepare for the first FDA audit, and the cost is likely to be in the region of $6000. If, as a result of the first audit by the FDA, the company's performance is deemed to be substandard, then a revisit is required, and the costs for this start at around $25 000.

The legislation regarding registration is outlined below:

'The Public Health Security and Bioterrorism Preparedness and Response Act of 2002 (the Bioterrorism Act) directs the Food and Drug Administration (FDA), as the food regulatory agency of the Department of Health and Human Services, to take steps to protect the public from a threatened or actual terrorist attack on the US food supply and other food-related emergencies.

To carry out certain provisions of the Bioterrorism Act, FDA established regulations requiring that:

- *Food facilities register with FDA, and*
- *FDA be given advance notice on shipments of imported food.*

These regulations became effective on December 12, 2003.

The FDA Food Safety Modernization Act (FSMA), enacted on 4 January 2011, amended section 415 of the Federal Food, Drug, and Cosmetic Act (FD&C Act), in relevant part, to require that facilities engaged in manufacturing, processing, packing, or holding food for consumption in the United States submit additional registration information to FDA, including an assurance that FDA will be permitted to inspect the facility at the times and in the manner permitted by the FD&C Act. Section 415 of the FD&C Act, as amended by FSMA, also requires food facilities required to register with FDA to renew such registrations every other year, and provides FDA with authority to suspend the registration of a food facility in certain circumstances. Specifically, if FDA determines that food manufactured, processed, packed, received, or held by a registered food facility has a reasonable probability of causing serious adverse health consequences or death to humans or animals, FDA may by order suspend the registration of a facility that:

1. Created, caused, or was otherwise responsible for such reasonable probability; or
2. Knew of, or had reason to know of, such reasonable probability; and packed, received, or held such food.'

5.5 AMAZON

Food Safety News ran an article in April 2018[6] about Amazon, the web host. It said:

'The 10-year-old dispute between the FDA and the world's largest Internet company by revenue crept into the headlines in only the past few days. Shevaun Brown, regional operations PR manager for Amazon Strategic Communications, does not see any problem with the mega-corporation refusing to register the food warehouse.

"Food and product safety are top priorities for Amazon and our fulfilment centres are not only permitted by state and local health departments, but we have a robust food safety programme to ensure our products are safe for our customers," Brown told Food Safety News on Monday. "The Amazon fulfilment centre at 1850 Mercer Road in Lexington, KY, is permitted in compliance with the Commonwealth of Kentucky."

Regardless of local or state laws, companies that transport and store food are subject to the federal Food Safety Modernization Act (FSMA), which seeks to prevent foodborne illness *instead of reacting to outbreaks. Federal registration of food warehouses began after 9/11 to defend against terrorism.*

But Amazon's Lexington warehouse is not among the more than 300 000 facilities registered with the FDA under the FSMA. More than 100 000 of those are warehouse or holding facilities.'

I think it is highly unlikely we will ever get harmony of legislation and therefore, with the advent of world-wide sales and supply through web hosts, for example, this will become more of an issue and harder to regulate. This will be compounded as local authorities find their budgets slashed further in the name of austerity, as outlined in the ironically named 'Financial Sustainability of Local Authorities 2018 report by the National Audit Office', which suggested that around 50% of their budgets had been stolen since 2010–2011.[7]

These differences in food regulations around the world, not just those outlined above but many more, lead to many RASFF notifications of non-compliant or adulterated foods. For example, there are 511 notifications relating to the banned use of chloramphenicol in crustaceans and honey.

5.6 ADULTERATED FOOD

The problem is that adulteration is difficult to define and means different things to different people. Adulteration was defined by Dr Henry Letheby, 1816–1876, as *'the act of debasing a pure or genuine commodity for pecuniary profit, by adding to it an inferior or spurious article, or by taking from it one or more of its constituents.'* To some (Caroline Walker, the distinguished nutritionalist, writer and campaigner and others), this includes: additives such as emulsifiers, synthetic flavourings and colours, which are added to make a food more desirable. Dr Henry Letheby's definition would seem, at least at first glance, to support her opinion. However, additives such as antioxidants (added to foods that contain fats to stop them going rancid), colours, emulsifiers, stabilisers, gelling agents and thickeners, flavourings, preservatives and sweeteners are permitted in foods and controlled by food law. In the eyes of the law, they are not adulterants. If they were classed as adulterants, then synthetic foods, such as an instant strawberry dessert that has never been near a strawberry (or any other fruit) or in fact anything containing additives, *i.e.* ready meals, sweets, snacks and numerous other products, would be considered adulterated, leaving the public with much less choice and eating only the most 'natural' ingredients (those grown or raised).

The public analyst's role is to help the courts decide what is adulterated and what is not; and, yes, the meaning of those words does change as research provides more evidence. For example, azo dyes (in some food colours), once permitted, are now considered harmful and have more recently been banned. Also, Sudan dyes, colourants used in oil solvents and polishes which were banned in the EU in July 2003, were found in chilli powder imported from India in 2005 and subsequently added to many foods, including pizzas, sauces and ready meals. Sudan dyes are now considered carcinogenic (an agent that promotes cancer), at least if present in sufficient quantities. Prior to 2003, they were a permitted additive. Another example is the 'Southampton Six': six colours linked by research at Southampton University with increasing ADHD (Attention Deficit Hyperactivity Disorder) behavioural problems in children. The colourings are Sunset yellow (E110), Quinoline yellow (E104), Carmoisine (E122), Allura red (E129), Tartrazine (E102) and Ponceau 4R (E124). Whilst there were arguments between researchers and EFSA as to the methodology chosen for this study, the EU has agreed that food and drink containing any of these six colours must carry a warning on the packaging. This will say '*May have an adverse effect on activity and attention in children*'.[8]

As a result, to enable action to be taken for a regulatory offence, enforcement officers have only to prove that a product was placed (made available on-the-market) which did not meet the legal requirements. It is for the defence to put forward, for example, a defence of due diligence, albeit on the lower standard of proof presented in a magistrates' court, namely on the balance of probabilities (the criminal courts require a higher burden of proof, namely beyond reasonable doubt). 'I didn't know', cannot be a defence; neither can you claim 'it was an accident'.

5.7 THE ELUSIVE DUE DILIGENCE DEFENCE

I believe that the first successful defence of due diligence for a food was accepted in court in April 1997.[9] Ms Henriques had a chocolate bar bought for her from a kiosk in Piccadilly Circus, London. Timothy Spencer, for Westminster Council, told Horseferry Road Magistrates Court, *'She partially unwrapped the chocolate bar, but noticed nothing unusual. She ate about three quarters of the bar and when she put the last piece into her mouth she bit on something hard. She saw a black, furry object surrounded by caramel and nuts. She thought it was a peach stone. She showed it to a colleague who reported seeing thick black fur and red stuff. She said it was a mouse.'* Working in conjunction with the Natural History Museum, Alan Parker (the Westminster Public Analyst), concluded that the furry object was the head and shoulders of a mouse which had a red colour deposited on its teeth (no analysis of the red dye was undertaken).

The prosecution alleged that the rodent had entered the manufacturing process in the UK and therefore the manufacturer was negligent. Mr Spencer informed the court that the manufacturer allegedly had a rodent problem at the time of manufacture (17 February 1997) and that more than 80 mice were found at the time by pest control officers.

Mr Meyer, a rodent control officer for the manufacturer, informed the court that the company had previously had infestations but that this particular mouse had, in his opinion, been brought into the manufacturing process when imported Turkish nuts were added to the confectionery. His evidence was based upon the staining on the teeth of the mouse, which was thought to be due to red rodenticide, as used in Turkey, as opposed to blue rodenticide used in the UK by the manufacturer. It was also claimed that the manufacturer was diligent in that it had regularly checked the importer. As a result, Stipendiary Magistrate Tim Workman ruled that the manufacturer could not be held responsible as he favoured that the offending mouse parts in the chocolate had been imported, and so the manufacturer could not be found guilty of negligence under the Food Safety Act 1990.

Under the Food (Safety) Act 1990, it is a defence for the person charged with an offence under the Act to prove on the balance of probabilities:

- that he took all reasonable precautions; and
- exercised all due diligence to avoid the commission of the offence by himself or by a person under his control.

It is the word 'all' in both points that leads to a nagging doubt in my mind.

The main part of my career involved working for a large local authority in their scientific service (public analyst). As part of my role, I established due diligence monitoring of our in-house catering organisation. We supplied around 10 million meals per annum to people in our care in around 600 establishments. The due diligence scheme was probably the only one of its kind in local authorities across the UK. In short, this included factory audits of suppliers as well as sampling and testing of raw materials (from suppliers) and final meals (in around 10 establishments on an annual basis) using both microbiological and chemical testing.

Latterly, as Head of Service, I was responsible for ensuring the council obtained good value for money. Consequently, every year (if it wasn't every year, it felt that frequent), we debated how much testing and monitoring we should undertake and how much of the tax payers' money we should use. We bought in good faith from trusted suppliers, so should we bother to undertake our own monitoring, let alone chemical and microbiological analysis? To solve this conundrum, I sought advice from colleagues in other local authorities, although I never found another who undertook this sort of monitoring. I asked friends in the large retail and food businesses who did undertake monitoring *etc.* but they couldn't give me a yardstick. I also consulted with enforcement colleagues and the regulator yet still could not establish a yardstick or clear definition of what 'all reasonable precautions' or what 'due diligence' actually meant.

The salutary lessons from the very sad death of a five-year-old boy in South Wales, who died after eating meat which had been contaminated with *Escherichia coli O157* following the poor practice of a butcher who supplied meat for school meals throughout the region,[10] combined with Professor Pennington's recommendations following his investigation into this tragic event and his subsequent comments regarding monitoring and funding,[11] all hardened our resolve to continue monitoring. We agreed that we would reduce our efforts if no problems were identified. However, we did occasionally find problems, sometimes within our processes, including, for example, small serving sizes (nutritional values are important in school meals) and occasionally with the products supplied. As a result, we continued to monitor at the same levels and also question whether we were doing too much.

No harm came to anyone in our care from eating contaminated food but if, God forbid, it had arisen, what would the public expectation be? That question brings me back to some unrelated research carried out by trading standards colleagues who asked how often the public thought petrol pumps were checked by enforcement officers to ensure they delivered the correct volume. The answer: at least monthly. The truth: at best annually. Perhaps, in the event of a tragedy, public expectation may not match the true situation. On my way into court once, I was told by the defending barrister that his simple definition of negligence was *'to not try'*. He went on to say: *'The defendant may be incompetent, but if he tried he couldn't be found negligent.'* Perhaps the simplest definition of due diligence is trying to check that the food is safe.

I am confident that my advice was reasonable. We were diligent (we probably did more than anyone else); we identified key control points; we learned from our monitoring and ensured it was dynamic and responsive. I passionately defended the level of monitoring every year and, to the credit of the council, we carried on monitoring at those levels even in times of austerity. However, I still struggle with the two questions: did we do too much or did we do enough?

A definition of due diligence remains elusive; I guess the jury is still out. Literally, it really is up to a court to say whether *'all'* reasonable precautions and due diligence have been exercised. I am aware of a case coming to court in the future where exactly this question may be asked. I will be watching to see what happens as then perhaps we will all have a clearer picture, but I am not holding my breath.

5.8 A NEW DAWN IN THE UK

EU law EC 882:2002 has arranged food enforcement in the same manner through member states. However, one major difference has been that the option to pursue food fraud has not been taken by the UK as it involves police and other food enforcement officers working together, and this has not been a priority for either in terms of food fraud. Neither has the law been particularly clear in this regard. It basically hasn't been on the horizon until Chris Elliott did his investigations and recommended a change.

Now in the UK, rather than Europe, things have changed. *Mens Rea* has been introduced to food enforcement: that is the intent to defraud. In Europe, the option to pursue a criminal case for food fraud has been available for many years and apparently has been followed, whilst in the UK, it has been 'available' but it has not been pursued. Food enforcement officers will need to be mindful of this when investigating, and clarity of working arrangements between those in the crime unit wishing to prosecute and colleagues undertaking market surveillance is required. Processes will also need to be clear as to how local officers pass this on to the food crime team for further investigation. The horsemeat case certainly enabled colleagues to 'cut their teeth' and bring about a successful prosecution, and I suspect the Farmbox case helped in this regard too. Seems the loveable rogue was in the wrong place at the wrong time, although I am not sure we like loveable rogues any more – that went in the 1990s.

5.9 CLARITY FOR THOSE INVOLVED IN PROSECUTING

What is needed to ensure food fraud prosecutions are successful in future?

It is likely that an operation will follow intelligence or market surveillance by local enforcement officers and may therefore not be initiated by the national food crime unit. Where this occurs, and fraud is expected, good liaison will be needed between the local and national teams. The national team will take over the administration of the prosecution and liaise with City of London Police and others, including the Crown Prosecution Service. I am interested to see if other police forces become involved in food fraud cases or whether the City of London Police remains the only force to investigate and prosecute.

Before this occurs, local enforcement officers need to be trained in the following: seizing relevant evidence and handling unused material, as well as being able to handle, search and disclose copious quantities of digital material extracted from computers *etc.* Whilst the local officers will hand over the issue to national colleagues, it is imperative that they are trained and competent in the above to ensure a smooth handover. LA trading standards officers will have previous experience in these areas from investigations into other non-food issues. Nationally, the Food Crime Unit and City of London Police officers will need to ensure there is early liaison with prosecutors and an good focus on case management issues such as identifying likely witnesses, unused material, seizing evidence, involvement of other parties, for example HMRC (Her Majesty's Revenue and Customs), search warrants restraint and what needs to be proven *etc.*

I have one concern if many fraud cases are identified and that is will the food crime team be able to cope with the workload and will the City of London Police have the available resources, but I guess that remains to be seen and tested. If predictions regarding food crime levels are correct then there will be many cases.

5.10 IN SUMMARY

Food safety law reflects society. It has become increasingly important to the public from the early days of the 19th century when scientists had 'celebrity status', giving sell-out lectures to the public on how to identify food fraud/adulteration (back then the terms were interchangeable) and to name and shame the criminals. Eventually, as the 1860 Food Safety Act and subsequent revisions became more successful in fighting crime, society relaxed and the popularity of food scientists sadly waned. Then, the nature, substance and quality, as defined in the Acts, became more important and this needed an 'average man' to determine what this meant in practice so the courts could enforce. The average man was designated as the 'Man on the Clapham Omnibus', a supposed ordinary and reasonable person who arrived, at least metaphorically, in our courts. He has been with us for around 100 years and is still the person to be considered by judges and all involved when considering what is reasonable, *i.e.* the nature, substance and quality so demanded by the consumer.

In my experience, politicians have been excellent arbiters of public opinion and what that man on the Clapham Omnibus feels. When they get it wrong, it results in a loss of their job at re-election time so they need to develop these skills. Government departments regularly consult politicians at all levels to decide strategy and help determine public opinion. The politicians, on behalf of the public, have decreed that 'heads will roll' and stiffer penalties should be available to the courts for the new crime of food fraud. Consequently, food crime has moved up the hierarchy of courts and is now on the A-list, the criminal arena: the Crown Court, not the entry-level Magistrates Court. The latter is for the 'regulatory offence' not the criminal offence of food fraud. Now, large corporations are being given larger fines. 'A zero has been added' to the fines being handed out by the courts and custodial sentences are available for perpetrators. Society has spoken, and there is now a cogent deterrent to the would-be fraudster. Perhaps this is exemplified by some recent sentences handed out by the courts:

5.10.1 ASDA

ASDA was fined £664 000 for cheese rolls covered in mouse droppings.[12]

5.10.2 Sideras, Nielsen and Ostler-Beech

Sideras was jailed for four years and six months and Nielsen for three years and six months, while Ostler-Beech was given an 18-month suspended sentence and a 120-hour community service order.[13]

5.10.3 Boddy and Moss

Moss received a four-month prison sentence, suspended for two years, and was ordered to pay costs of £10 442, payable within six months.

Boddy was fined a total of £8000 and ordered to pay costs of £10 442, payable within six months.

5.10.4 Zaman

Mohammed Khalique Zaman, a reckless and cavalier restaurant owner, was found guilty of the manslaughter of Mr Wilson and six charges of contravening various food safety requirements. He cut corners by using cheaper ingredients containing peanut and gave them to a customer who had declared his allergy to peanut. The take-away food provided was labelled as nut free, but wasn't.

He was sentenced to six years' imprisonment for the manslaughter and up to nine months for each of the food safety offences, all concurrent.[14]

REFERENCES

1. Horsemeat scandal: David Cameron says offenders will feel full force of law, https://www.theguardian.com/uk/2013/feb/13/horsemeat-scandal-david-cameron-food, [accessed August 2018].
2. Food crime unit to be setup in wake of horsemeat scandal, https://www.channel4.com/news/food-crime-unit-to-be-setup-in-wake-of-horsemeat-scandal, [accessed August 2018].
3. John Barnes on Sentencing Guidelines, https://shieldsafety.co.uk/john-barnes-on-sentencing-guidelines, [accessed August 2018].
4. Regulation (EC) No 178/2002 Of The European Parliament And Of The Council of 28 January 2002, https://eur-lex.europa.eu/legal-content/EN/TXT/PDF/?uri=CELEX:32002R0178&from=EN, [accessed August 2018].
5. Food safety in the US-EU TTIP negotiations, http://capreform.eu/food-safety-in-the-us-eu-ttip-negotiations/, [accessed August 2018].
6. Amazon says food warehouse isn't subject to FDA food rule, http://www.foodsafetynews.com/2018/04/amazon-says-food-warehouse-isnt-subject-to-fda-food-rule/#.WyQ8TqdKjIU, [accessed August 2018].
7. Financial sustainability of local authorities 2018, https://www.nao.org.uk/press-release/financial-sustainability-of-local-authorities-2018/, [accessed August 2018].
8. Food Additives, https://www.food.gov.uk/safety-hygiene/food-additives, [accessed August 2018].
9. Mouse 'found in Topic bar', https://www.independent.co.uk/news/mouse-found-in-topic-bar-1179399.html, [accessed August 2018].
10. Boy, 5, dies in E.coli outbreak, http://news.bbc.co.uk/1/hi/wales/4307442.stm, [accessed August 2018].

11. The Public Inquiry into the September 2005 Outbreak of *E.coli* O157 in South Wales, http://www.reading.ac.uk/foodlaw/pdf/uk-09005-ecoli-report-summary.pdf, [accessed August 2018].
12. Asda is fined £664, 000 after customer eats cheese rolls covered in MOUSE droppings forcing inspectors to close infested bakery for 10 days, http://www.dailymail.co.uk/news/article-3912936/Asda-fined-664-000-customer-eats-cheese-rolls-covered-MOUSE-droppings-forcing-inspectors-close-infested-bakery-10-days.html, [accessed August 2018].
13. 'Greedy' horsemeat fraudsters sent to prison, http://www.horseandhound.co.uk/news/horsemeat-fraud-jailed-628057#cD4tuod6KDvovXdt.99, [accessed August 2018].
14. Peanut curry death: Restaurant owner Mohammed Zaman jailed, http://www.bbc.co.uk/news/uk-england-36360111, [accessed August 2018].

CHAPTER 6

Mission Impossible; Regulating a Global Marketplace

In this section, I shall use the terms Internet and World Wide Web (web) as if they are interchangeable. I know they are not, and I hope you'll forgive me. Systems like instant messaging use the Internet and other systems of providing information use the web to present pages of information to the user. Both systems are having a significant impact on how we procure food and, I believe, will revolutionise our high street.

The Internet has changed the way we shop; mobile systems such as tablets and smartphones enable us to shop anywhere, anytime. The smartest retailers are adapting and learning the best ways to use these new channels to get products before consumers in a timely way. To facilitate 24-hour shopping, marketplaces such as Amazon, YouTube and eBay, to name a few, have developed significant numbers of suppliers of all sorts of goods from around the world. The unique selling proposition being cheap price (some even show price comparison) fast delivery and the convenience of shopping whenever the customer wants. Consequently, a world-wide supply of goods is available year-round and delivered to your door in double quick time. EU law allows free passage of goods across all member states, but the same isn't true for products delivered from outside the EU, as they must be brought to the attention of Border Control before

---

The Horse Who Came to Dinner: The First Criminal Case of Food Fraud
By Glenn Taylor
© Glenn Taylor 2019
Published by the Royal Society of Chemistry, www.rsc.org

being allowed to enter the EU. Delivering directly to your front door may have the potential to circumnavigate Border Control systems (Customs) if, for example, the goods are posted directly. To preserve their reputation, the web marketplaces will need to ensure that the sellers on their sites undertake the necessary checks to ensure compliance with legislation, *i.e.* the traceability of produce and testing for compliance when necessary. This will not always be easy and, with a focus on costs, will the suppliers try to reduce the cost of testing and traceability? Many large high street retailers have ceased trading, citing Internet suppliers as one of the causes of their demise. I expect web sales to continue to grow and fear that the high street may continue to flounder. This is the new transition economy; it is no longer just the BRICS (Brazil, Russia India, China and South Africa) economies that EU enforcement and the regulators will need to watch, particularly whilst they establish their own enforcement. There will be a world-wide supply of goods, possibly from small fast-developing suppliers in other growing economies, which may not be closely scrutinised by local enforcement, and the products may not pass through EU border inspection posts and therefore may go unchecked.

Traditionally, the web has not been used for food sales, although there are signs of that changing as more retailers and marketplaces are offering delivery of food. Our research into food supplements sheds much light on the potential issues for web-based supply and the challenges that enforcement might face as food supply migrates to web-based sales. Our first published paper in this area was called 'Mission Impossible',[1] which looked at supplement usage and the difficulties in regulating web-based purchases. This led to more research, entitled 'Russian roulette',[2] into why users ignore the regulator's advice and listen to the 'wise counsel' written on spurious webpages of user groups and of gym buddies, even when they know it can, and does, result in death. The findings shocked me as I couldn't believe the risks consumers would take when buying from unchecked sources over the web. Having discovered this in food-related products, I wonder if this could be the case for future food purchasing as we migrate from the high street. If this is the case, then how could we regulate food purchasing? With no world-wide harmony of legislation, what is permitted in one country may not be permitted in the EU or England post-Brexit.

Dark clouds are on the horizon for food and food products' regulation if we fail to adapt to changing habits of consumers and ignore the decision processes they use in making these purchases.

6.1 FOOD SUPPLEMENTS: 'THE GYM BUNNY'S' LITTLE HELPER!

This is an area that has worried me for some time. Epidemics in obesity have been documented in many developed countries, and the desire for a well-toned body and instantaneous results has driven many to seek help from chemistry in the form of a supplement that will produce rapid results in the quest for the perfect body. 'Supplements' cover a wide range of products, some of which err towards medicinal compounds used by athletes who seek help with their training from 'magic substances', which can include questionable ingredients such as steroids and other banned substances. As a result of these quasi-medicinal claims, the products fall between two areas of regulation: food supplements and medicine.

6.1.1 Definitions for Medicines and Supplements

Medicines must be licensed by the appropriate body; in the UK the Medicines and Healthcare Products Regulatory Agency (MHRA).

Supplements is a complex area of EU law. In a nutshell, food supplements are intended to supplement the diet – for example, vitamins and minerals – and they should have a nutritional and/ or physiological benefit. EU legislation does permit other substances to be referred to as supplements – for example, proteins. There is confusion within the EU as some member states declare a product – for example, some herbal substances – a medicine while others call it a supplement, and if that's not bad enough, some web-based sellers have provided products that are unlicensed and neither a supplement nor a medicine, but claim benefits and offer them for human consumption. EU guidance states: *'Food business operators market food supplements, which are concentrated sources of nutrients (or other substances) with a nutritional or physiological effect. Such food supplements can be marketed in "dose" form, such as pills, tablets, capsules, liquids in measured doses etc.*'³ Some supplements have been sold as not for consumption, the same system used for legal-highs, despite the seller being well aware that they were going to be taken by the consumer. Legislation within the EU is available to stop the sale of chemicals which are sold on the basis of false or misleading health claims, especially those products that are unsafe. However, this is not the same across the globe, and a quirk in the law in the United States allows ingredients which were established before 1994 to be marketed without evidence of their effectiveness or safety.[4] Before you lose the will to read on, let's just live with abuse of the term supplement by many sellers, particularly when the substance has very limited information in the way of an efficacy trial proving it works or that it meets safety standards.

The world-wide sports nutrition market has grown rapidly since 2000 and is predicted to become worth around $40 billion per year by 2020.[5] Sales and marketing by this industry now targets the recreational and lifestyle users. Not all products should cause concern as many are from reputable organisations, but some of these quasi-medicines, sometimes referred to as nutraceuticals, are distributed and marketed *via* the web and occasionally passed on by an acquaintance at the local gym; this makes enforcement far more difficult.

6.1.2 Is UK Enforcement Doing Enough?

Hazards relating to concentration, composition, individual contaminants and supplement interactions present an increasing public health risk. Approximately 25% of sports supplements are not entirely as declared on the label, with anabolic steroids being the most common adulterant.[6] Notifications logged in the EU's RASFF database through imports or market surveillance are typically logged for poor quality control issues. These notifications have steadily increased six-fold for supplements in the past 10 years, with the United States and China being the major transgressors (Figure 6.1). Finland and Italy lead in detections, mainly notifying unpermitted substances and contaminants in sexual-enhancing or weight-loss supplements.[1]

Purchasers of supplements perceive these quasi-medicinal products to have a beneficial impact on performance and health. Supplement users may consume poorly controlled

Mission Impossible; Regulating a Global Marketplace

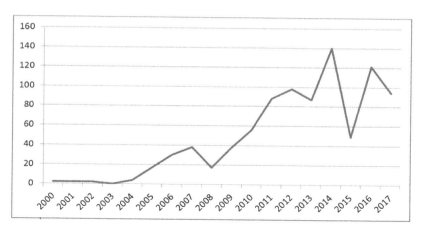

Figure 6.1 The increasing number of notifications logged per year (2000–2017) in the EU Rapid Alert System for Food and Feed (RASFF).

substances with little or no clinical guidance and a paucity of factual data on the contents, concentrations, doses or effects of the substances they take. In the absence of controlled clinical trials, adverse health effects from supplements can be assessed through epidemiological studies using patients and doctors' reports when the key ingredient is as the label declares and is therefore known to the subject or practitioner. However, when unknown ingredients are involved, *i.e.* undeclared contaminants, then any observed side effects may be attributed to the declared ingredients when they may, in fact, be caused by the contaminating substance. Therefore, unless the supplement itself is tested, side effects can be inaccurately attributed to particular supplements, leading to incorrect policies and/or warnings. Regulation and surveillance of dietary botanical supplement use is lacking in most countries, with calls for improvement in the United States.[7]

The food supplement market in the UK is worth around £0.5 billion, which equates to around 0.7% of the UK food market. Enforcement officers submit around 50 000 samples each year to their public analyst laboratories for chemical analysis, but only 0.2% of these samples relate to supplements. An analysis of the RASFF notifications indicates poor quality control: contamination with industrial chemicals such as dioxins and polyaromatic hydrocarbons and the use of unsubstantiated health promoting ingredients – for example, hormones or steroids. Therefore, these products deserve proportionately more attention than food, not less as is currently the case. The number of detections recorded in the RASFF database rose dramatically after the introduction of the 2002 Directive[8] demonstrating the need for an increased vigour by enforcement on this key issue. However, the number of RASFF notices in proportion to the number of samples, and the importance of the issues noted, warrant these products remaining high on the list of priorities for enforcement. The exponential increase in the numbers of RASFF notifications suggests that, however these products are regulated (food or medicine), there remains a clear need for citizens to be protected from hazards present in supplements by rigorous enforcement; enforcement that ensures clarity of labelling, appropriate quality control and appropriate guidance for the unsuspecting consumer, however they purchase the product – an enormous challenge and one that currently receives insufficient attention.

As if to demonstrate the problem, in 2013 the use of a magic diet pill sold over the Internet resulted in six deaths in the UK and around 60 world-wide. I became involved in trying to ban the use of the chemical 2,4-dinitrophenol (DNP) as a diet pill in the UK.

DNP was sold as 'the dream diet pill'; the tablet that resulted in weight loss without dieting. Despite all the deaths, it is still recommended by weightlifters who want a more defined six-pack. The hazardous chemical was never designed as a diet pill or medication in the first place. It was originally mixed with picric acid by French munitions workers to make First World War explosives. The workers noted that they lost weight and felt tired and over-heated. In 1930, it was suggested by Cutting *et al.*, that it could be used as a fat-burner/diet pill due to its ability to '*markedly augment the metabolism*', leading to rapid weight loss of up to 1.5 kg (3.5 pounds) per week by literally burning fat and carbohydrates. DNP use gained in popularity during the 1930s but significant variations in the way people responded to the chemical were noted along with more side effects, and many deaths occurred before precautions were taken. This information led to research by Tainter and Cutting in 1935 and Loomis and Lipmann in 1948,[9] which concluded that, when this compound was ingested, cell mitochondria are short-circuited, leading to fat and carbohydrate stores not being metabolised into useful energy but producing heat, which will result in a loss of weight and may cook the user to death. In 1938, the compound was declared extremely dangerous and not fit for human consumption by the Federal Food, Drug and Cosmetic Act 1938.[10] The key problem is that an effective dose has not been proven by a proper clinical trial because the compound is too dangerous. However, some users, ironically weightlifters and other athletes, have argued that the use of small dosages will safely result in weight loss. Hence our investigations into the use of the compound and others like this which resulted in the research paper 'Russian roulette with unlicensed fat-burner drug 2,4-dinitrophenol (DNP): evidence from a multidisciplinary study of the Internet, bodybuilding supplements and DNP users'.

The RASFF,[11] a communication system used by enforcement across the EU, shows that the Finnish authorities alerted fellow countries to the danger having found it in capsules in June 2003. Selling this compound as a diet aid contravenes EU safety laws. We had many discussions with the authorities as to how to regulate and who should take the lead. Were these compounds:

- medicines – no, they were not licensed and not purporting to be a medicine;
- a food – no; or,
- a food supplement – no, see above.

This caused enormous difficulty as no organisation had the responsibility for dealing with this issue. To their credit, senior staff at the FSA were the only government body willing to step up to the plate and lead a regulation initiative to halt the sales of these compounds, eventually resulting in a 'ban' in October 2013. For the record, it is practically impossible to ban a chemical from sale, only to ban the sale of something for a particular use. Despite knowing the issues, DNP might still be purchased from unscrupulous suppliers in other countries, as reported in the *Daily Mail* at the time of one of the deaths from this compound.[12] Here, the sellers were not selling the chemical in the normal manner but transporting it in capsules, which in the opinion of many, were simply aimed at people who wanted to ignore the advice and ban. The *Mail* reported that they even offered to disguise the import paperwork so that the produce could get past customs. Since then, the police, FSA and local authorities have worked together to stamp out the illegal sale of DNP, checking high street sellers and stopping Internet sales by closing down websites and reminding companies of the criminal sanctions available to the courts to deal with any person or company found to be supplying DNP for public consumption.

Our research[2] into online DNP sales, at the time of the ban, showed that there is often a mismatch between the locations of the 'IP addresses' and shipment information, indicating that DNP suppliers hide behind layers of webpages and use multiple websites simultaneously, making policing more difficult. Whilst it was sold on some sites with health warnings, elsewhere it was marketed as an effective fat burner. Our investigations concurred with the *Daily Mail* in that the suppliers openly offered to help get around the law using false labels *etc.* One typical example was:

> "We have a success rate of 98% at USA, Canada and European customs with Registered Mail Method only. Our company eliminates the remaining 2% risk and gives you a customs seize guarantee. If your order is seized at the customs, we ship one more time for free. Seized orders must be sent to a different address because customs may flag your address. And the best part is you do not have to pay the shipping charge again."

Customers were given a plethora of advice as to how to take the compound with suggestions for other compounds to assist, such as diet suppressors like insulin and thyroxine, antihistamines to manage the allergic reaction and caffeine or Kratom to counterbalance the lethargy often noted when taking DNP.

Sadly, death is not restricted to DNP. We investigated the death of another athlete who used several compounds in an attempt to boost his performance, and one of these was Kratom (*Mitragynine speciosa*), a compound supplied to workers in paddy fields in Vietnam to help them stay working for longer in hot, humid and difficult conditions. It seems that many will ignore regulation and advice from qualified practitioners and circumnavigate the 'regulated high street' and 'web companies', preferring to use unreliable information and suppliers, even where death may be the result. Respondents told us that where a short-term gain was required, for example to fit a dress for a wedding or look good in a competition, then the risk was worth it. Clearly, respondents don't expect to die from these compounds and they feel the advice they receive from websites *etc.* is sufficient to avoid that outcome, but that clearly wasn't the case for everyone.

6.2   SILDENAFIL CITRAT

This is one issue that intrigues me. Sildenafil citrate has a trade name: Viagra®. It is a vasodilator that will enable increased blood flow around the body. The 'side effect' noted during efficacy trials was the impact on erectile dysfunction. Since this was discovered, it has been patented for use in this way and, in the UK, is a licensed medicine, regulated by the MHRA, which must be prescribed by a GP. However, there are other uses. For example, it has been used by athletes to increase blood flow and thus increase the effectiveness of hard workouts in the gym. Increased oxygen transfer enables athletes to work out for longer and the effect is a faster build-up of muscle. It has also been taken by users of illicit drugs which may lead to erectile dysfunction, for example MDMA (Ecstasy).

There have been 104 RASFF notifications, 102 in dietetic foods, one in confectionery and another in a tea, and the frequency of detection has increased over the five years to 2016 (the last published RASFF report). So why is Viagra being found at Border Control across the EU? Is it for its original purpose, with people bypassing medical checks, or is it for athletes to use for another purpose? Analogues of sildenafil (compounds which are similar but not exactly the same, and thus not tested for efficacy or safety) have been found as adulterants in herbal aphrodisiac products.[13]

Many countries struggle with the issue of analogues of illicit drugs which were sold as legal highs. I was heavily involved in policing the sale of legal highs – compounds that mimic illegal drugs in terms of their chemical structure but, by virtue of not being identical to the illegal compound, were not illegal. Consequently, they were nicknamed 'legal highs', which suggested to

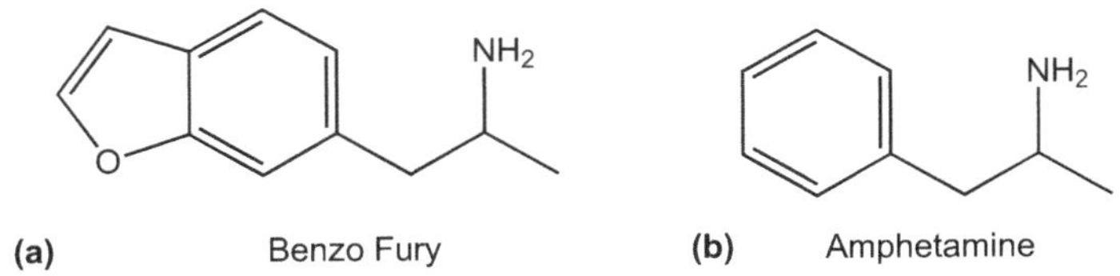

Figure 6.2 (Left) Benzo Fury chemical structure, an analogue of an illicit drug sold as a 'legal high'. (Right) Amphetamine, a class B drug and structural analogue of Benzo Fury.

some that they were legal to use and therefore somehow safe, or at least safer than an illegal compound. It seemed at the time that users thought if a compound wasn't illegal for use it must have been tested and declared safe. They were assumed to behave in a similar manner to the illegal substance; for example, the 'legal high' Benzo Fury and the class B drug Amphetamine – under the Misuse of Drugs Act 1971 – are predicted to behave in a similar manner on the user (Figure 6.2).

Until a change in the law, the products were sold in local headshops and imported *via* the web. Those we tested were often not as described and were provided by 'organisations' that had poor quality control and sold in differing concentrations. As a result, the user could not be sure of what they were buying. This sounds remarkably similar to the supplements outlined above. The UK eventually banned legal highs in 2016.[14] Analogues of prescription medicines have not yet received the attention of the authorities, so consequently they will be available on the web.

I was worried and amazed by the findings in our research that some people would take these compounds, accepting the 'advice' from the web even when that advice recommended the use of other compounds – some prescription only, and others illegal. It concerned me because current enforcement systems will simply not work in this environment. Ban the chemical by all means, and I was keen to do just that, but it will be made available from another source or an alternative (an analogue), which is likely to be as toxic.

6.3 SO WHAT DOES THIS HAVE TO DO WITH FOOD FRAUD?

A global food supply is necessary to satisfy the demands of consumers who desire a continuous, year-round, supply of seasonal products and access to foods traditionally served in other areas of the world. Some 50% of the food available for retail sale in the UK is sourced within the UK,[15] and approximately 30% is sourced from within the rest of the EU; the remaining 20% is imported from outside of the EU.[16] Yet 80% of the issues recorded on the RASFF database relate to the 20% of the foods imported from outside of the EU. Thus, just like the products referred to earlier, there is a world-wide supply; consequently, it is imperative that safety and security are world-wide priorities, and every country must ensure compliance and appropriate testing regimes in their territory. Within the EU, free movement of food across its borders is allowed based on the level of control imposed on food producers within the member states. However, given the different interpretations of food safety, global compliance is difficult to ensure as adherence in one country does not necessarily result in conformance with EU regulations. The Beijing Declaration,[17] signed by over 50 countries, resulted from the need to ensure food producers across the world adhere to strict legislative standards so that food produced in the signatory's country is safe. Thus, in theory, should the Beijing Declaration be adopted and implemented across the globe, transportation of (safe) food produce should be unlimited. If practice fails to match the theory, and it does on a regular basis, hence the RASFF notifications, then the border inspection post, the first port of call, is vital for our protection and ensuring that food is of the nature, standard and quality so demanded by the consumer, *i.e.* compliant with EU law. In the UK, financial responsibility for border inspection lies with local authorities and port health authorities. However, since 2003, the FSA (the Regulators), in conjunction with the Department of Health, has provided around £5.4 million to local authorities and port health authorities[15] to enable additional surveillance of imported food, in recognition of the importance of food enforcement at the ports. This additional funding stopped in 2015.

Border inspection is not a barrier to trade, but a necessary part of food and feed enforcement, which is supplemented by other actions, such as market surveillance and the additional focus from the regulators. This places an unequal burden on member states, particularly those without border inspection points. However, it need not be so important if all signatories of the

Beijing Declaration manage to ensure the viability of their own systems and if more members sign up to the agreement. Given the levels of non-conformance noted by the EU, not all signatories to the Beijing Declaration have sufficient monitoring systems in place which can guarantee that produce from their country will meet the requirements of the EU. Without world-wide harmonisation of legislation – for example, in the area of permitted additives – there will always be the potential for differences in additives found in food, some of which may be permitted in the exporting country and not permitted in the importing nation. The legislation states that additional import restrictions are appropriate to protect human health and that there is justification in terms of scientific-based evidence. The levels of detections in some member states and the levels of non-compliance found as a result of additional focus from the regulatory bodies provide evidence to confirm that there is a solid foundation for continuing with border inspection and emergency controls, even if the additional funding is no longer available.

In our paper 'The Procrustean bed of EU Food Enforcement'[18] we sought to assess how well member states protected each other across the EU. For each member state, we compared import tonnage of food, population and border inspection notifications made, as recorded on the RASFF. Comparison of import tonnage with population shows that some member states import far more food than others. The Netherlands was one example, and Rotterdam is clearly a gateway port to the EU. Wide variations in food safety practice exist between member states. Variations include both the number and type of contributions to the RASFF database, with some member states being relatively highly active in the key class of Border Notifications. An evaluation of import tonnage per border notification revealed considerable differences, up to 129-fold, within member states. Some member states punched above their weight, Malta being one example, and three led the way – namely Italy, Germany and the UK (in decreasing order). Some of the gateway member states reported lower than average numbers of notifications of non-compliance, and that is worrying, as food imported through these ports is free to travel across the EU without further inspection.

Figure 6.3 shows the relative plots. Graph A, for example, shows food import tonnage *versus* population, and it can be seen

Mission Impossible; Regulating a Global Marketplace 115

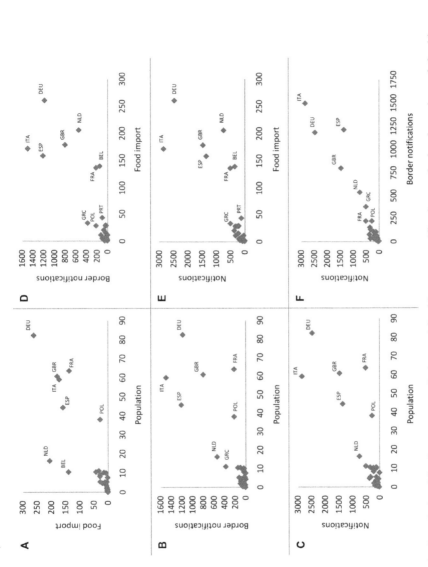

Figure 6.3 Pairwise relationships between: population and food imports, notifications at the border or in total; food imports and notifications; both at the border and in total; and total notifications and border notifications. Reproduced from ref. 18 with permission from Elsevier, Copyright 2013.

that The Netherlands' (NLD) imports are significantly higher compared to its population when compared with other EU member states, whereas Polish (POL) imports are relatively low.

Given that the EU emergency control and imported food safety strategy is only as strong as its weakest link, a counter-measure to inconsistency or weakness at Border Control might be to ensure that a product complies with EU legislation at source. Whilst the principle of monitoring at source is laudable, *i.e.* the Beijing Declaration, food and feed enforcement will never be a Procrustean bed where one size fits all in terms of protection, and the EU is correct to promote a mixture of measures necessary to protect the health of citizens, including a vibrant Border Control system, local policing and monitoring at source.

Our research covered the period 2003–2008, the time before the financial crisis. Since that time, austerity has continued across the EU. With increasing austerity, the situation is not likely to improve, and since the publication of this research in 2013, many, if not all, member states have cut resources on food protection. So, in a nutshell, the key defence in food safety across the EU is variable, to say the least.

David E. Gumpert reported in the *San Francisco Chronicle*:

'The hopeful news in all this is that in the process of creating so much toxicity both the distressed loans and the distressed food are teaching us important lessons about the limits of scale and regulation that support the massive globalisation of the last decade. We are learning that regulators have lost the ability, if they ever had it, to truly monitor the extent of the problem.'[19]

It seems the issue isn't unique to the EU. Currently, the EU adds an import levy of 17% on foods; I hope that this money is used to fund border inspection, although I doubt it. It has been suggested that post-Brexit the UK will scrap this levy to provide cheaper food for the consumer. Is that wise? We have already seen falling investment in our enforcement. Now couple this with a new channel, *i.e.* food supplied direct from the supplier from across the globe *via* the web, possibly bypassing border inspection, and the words reported in the *Daily Mail* that sellers will alter import paperwork to help bypass the law in the UK come to mind. I know this related to DNP, a supplement not a food, but if they are willing to do that where a death may occur, is it not conceivable that the law will be flouted for a sausage, or indeed any food where a good profit can be made? After all, it happened to food targeted at our most precious and most vulnerable, *i.e.* infant formula, and there are many other food scandals from across the globe that demonstrate this willingness to flout the law.

The following quote from Bee Wilson, which was published in *The New York Times* regarding melamine in baby milk, is pertinent to the fast growth of Internet-based sales.

> '*The similarities between China today and New York 150 years ago [I'd like to add London and Paris] shouldn't come as a great surprise. Adulteration on such a scandalous scale occurs in societies with a toxic combination of characteristics: a fast-growing capitalist economy coupled with a government unable or unwilling to regulate the food supply. In such get-rich-quick societies, there is a huge temptation to tamper with food, particularly when margins are low. The rewards are instant, and it's not always easy for consumers to detect the difference between the pure and the doctored—particularly with a substance like milk, which we have been taught to trust implicitly.*'[20]

I fear for the future of food enforcement and food regulators, as both struggle now, but as the Internet becomes more prominently used by consumers, those with a responsibility for ensuring compliance must work with the Internet providers and web hosts. Governments must ensure sufficient resources are made available to enforcement and find other ways of getting messages across to people who will ignore advice from official sources in the desire to obtain a quick fix to a problem, or a cheaper alternative.

REFERENCES

1. A. Petroczi, G. Taylor and D. P. Naughton, *Food Chem. Toxicol.*, 2011, **49**(2), 393–402.
2. A. Petróczi, J. A. V. Ocampo, I. Shah, C. Jenkinson, R. New, R. A. James, G. Taylor and D. P. Naughton, *Substance Abuse Treatment, Prevention, and Policy*, 2015, vol. 10, p. 39.

3. Food supplements, https://ec.europa.eu/food/safety/labelling_nutrition/supplements_en, [accessed August 2018].
4. Dietary Supplement Health and Education Act of 1994, https://ods.od.nih.gov/About/DSHEA_Wording.aspx, [accessed August 2018].
5. Global Sports Nutrition Market Size, Share, Development, Growth and Demand Forecast to 2022 – Industry Insights by Type and Distribution Channel, https://www.psmarketresearch.com/market-analysis/sports-nutrition-market, [accessed August 2018].
6. M. R. Cole and C. W. Fetrow, *Am. J. Health-Syst. Pharm.*, 2003, **60**(15), 1576–1580.
7. R. B. Wallace, B. M. Gryzlak, M. B. Zimmerman and N. L. Nisly, *Annals Pharmacother.*, 2008, **42**(5), 653–660.
8. Directive 2002/46/EC of the European Parliament and of the Council of 10 June 2002 on the approximation of the laws of the Member States relating to food supplements, https://eur-lex.europa.eu/legal-content/EN/ALL/?uri=CELEX:32002L0046, [accessed August 2018].
9. M. L. Tainter, W. C. Cutting and A. B. Stockton, *Am. J. Public Health Nations Health*, 1934, **24**(10), 1045–1053.
10. J. Grundlingh, P. I. Dargan, M. El-Zanfaly and D. M. Wood, *J. Med. Toxicol.*, 2011, 7, 205.
11. RASFF – Food and Feed Safety Alerts Website, https://ec.europa.eu/food/safety/rasff_en, [accessed August 2018].
12. 'No better than cocaine pushers': MailOnline exposes unscrupulous Chinese factories selling toxic DNP diet pills that have killed six Britons and 60 worldwide by 'burning them to death from inside', http://www.dailymail.co.uk/news/article-3078573/They-no-better-cocaine-pushers-DNP-victim-Eloise-Parry-mum-s-fury-MailOnline-exposes-unscrupulous-Chinese-factories-selling-deadly-diet-pills-burned-daughter-inside.html [accessed August 2018].
13. B. J. Venhuis and D. de Kaste, *J. Pharm. Biomed. Anal.*, 2012, **69**, 196–208.
14. Collection, Psychoactive Substances Act 2016, https://www.gov.uk/government/collections/psychoactive-substances-bill-2015 [accessed August 2018].
15. Food Standards Agency Annual Report and Consolidated Accounts 2011/12, https://assets.publishing.service.gov.uk/

government/uploads/system/uploads/attachment_data/file/247087/0036.pdf [accessed August 2018].
16. Food: an analysis of the issues (January 2008), http://webarchive.nationalarchives.gov.uk/+/http:/www.cabinetoffice.gov.uk/media/cabinetoffice/strategy/assets/food/food_analysis.pdf [accessed August 2018].
17. Beijing Declaration on Food Safety, http://www.wpro.who.int/foodsafety/documents/beijing_declaration.pdf [accessed August 2018].
18. G. Taylor, A. Petróczi, T. Nepusz and D. P. Naughton, *Food Chem. Toxicol.*, 2013, **56**, 411–418.
19. What raw milk and the economic meltdown have in common, https://www.sfgate.com/opinion/article/What-raw-milk-and-the-economic-meltdown-have-in-3190426.php [accessed August 2018].
20. The Swill is Gone, https://www.nytimes.com/2008/09/30/opinion/30wilson.html?=_r=6& [accessed August 2018].

CHAPTER 7

Thinking Like a Food Fraudster

7.1 OVERVIEW

Professor Elliott's review into the integrity and assurance of food supply networks challenged all involved to work differently and start 'thinking like a criminal' to better understand fraud. In response to that challenge, I was asked by the Food Law Enforcement Practitioners (FLEP) group to lead EU-wide research in conjunction with EU enforcement partners, academia and industry. At the outset, the research team debated the factors that lead people to choose a criminal lifestyle and the fact that fraudsters may simply look for any opportunity, irrespective of the market. After further consideration, it was decided to focus on food fraud alone, and therefore 'thinking like a food fraudster' became the mantra. We concentrated on activities that tricked or cheated consumers in order to profit through adulteration or misrepresentation and fraud. We split the research into three areas, the first sets the background to food fraud, the second highlights where a fraudster might attack and the third focusses on defence strategies.

7.2 LEGISLATION

Fraud in general is defined as deliberate deception intended to achieve financial or personal gain, but as yet, no separate legal

---

The Horse Who Came to Dinner: The First Criminal Case of Food Fraud
By Glenn Taylor
© Glenn Taylor 2019
Published by the Royal Society of Chemistry, www.rsc.org

definition for food fraud exists. Several have been proposed, and perhaps the best is suggested by John Spink:[1]

> 'a collective term used to encompass the deliberate and intentional substitution, addition (or dilution), tampering, or misrepresentation of food, food ingredients, or food packaging; or false or misleading statements made about a product, for economic gain.'

Since the Adulteration of Food Act 1860, food enforcement officers have relied on the terms 'adulteration' and 'substitution' because it is relatively straightforward to prove that a strict liability physical act (*Actus Reus*) has occurred, *i.e.* the food was adulterated or substituted. The penalties for adulteration under the Food Safety Act 1990[2] (the latest iteration of the 1860 Act) are lower than those for criminal acts, including fraud, leaving food enforcement out of step with other areas of criminal prosecution. A successful food fraud prosecution requires proof of two elements – a completed act (*Actus Reus*) as well as a dishonest intent (*Mens Rea*), with dishonest intent requiring knowledge and/or recklessness. There is inevitable overlap in the above definitions. A food which is rendered unsafe may arise from a fraudulent act; likewise, a food which is not of the expected nature, substance or quality may be due to fraud or because less than adequate systems are in place which, it still may be argued, increases the profit. If this is 'sold to the purchaser's prejudice', as inevitably it would be, then public perception is likely to be that fraudulent behaviour has occurred, even though that may not strictly be the case in law. Since Horsegate, there has been political support for the prosecution of perpetrators and the handing-down of larger sentences. This has led to a desire to prove fraud rather than simple adulteration or substitution. The key issue in this new climate will be ensuring that fraud is considered at the earliest stage of evidence-gathering as it requires specialist enforcement officers with appropriate training. Hence the need to set up a food fraud unit, as suggested in Professor Elliott's report.

The uninitiated could be forgiven for thinking that food fraud is a new phenomenon introduced after Horsegate. Sadly, this is not the case, as it can be traced back many centuries and probably from the beginning of bartering or selling food. There are several references to adulteration of wine and bread which can be traced back as far as the Greeks and Romans. Pliny[3] (AD70) recorded *'the wheat of Cyprus is swarthy and produces dark bread, for which reason it is generally mixed with white wheat of Alexandria'*. He noted, with disapproval, that some bakers kneaded their bread with seawater so that they could save on the cost of salt, and detailed how white earth 'Leucogee' (alum, associated more latterly with dementia) was added to bread to bulk its weight.

In the mid-1800s, two London-based scientists Fredrick Accum and Dr Arthur Hill Hassall, led the charge to stop food fraud. In Accum's book, *A Treatise on Adulterations of Food and Culinary Poisons*,[4] and through information released to the media, Accum and Hassall named and shamed London's food fraudsters. A look at Table 7.1, of fraud published at the time, is revealing, not least because it shows us likely areas of fraud we still encounter some 150 years later.

From Table 7.1, the olive oil which contained lead from presses was very unlikely to be a result of fraud.

Moving to more recent times, the RASFF annual report for 2012[5] provides data on foods which failed to meet compositional standards.

List in order of frequency of detection:

- Dietetic foods
- Fruits and veg
- Fish
- Cereals
- Cocoa, coffee and tea
- Herbs and spices
- Prepared foods and snacks
- Beverages
- Soups
- Fats and oils
- Meats
- Confectionery

Whilst the foods have not met EU compositional standards, this may not be the result of fraud as dishonest intent must also

Table 7.1 Examples of foods and adulterants in London in the mid-1800s, as reported by Accum and Hassall. Adapted from ref. 27 with permission from the Royal Society of Chemistry.

Food	Adulterant
Red cheese	Coloured with red lead (Pb_3O_4) and vermilion (mercury sulfide).
Confectionery	White comfits often included Cornish clay. Red sweets were coloured with vermilion and red lead. Green sweets often contained copper salts (verdigris: basic copper acetate; Scheele's: acidic copper arsenite; or emerald green: copper arsenite).
Olive oil	Often contained lead from the presses.
Vinegar	'Sharpened' with sulfuric acid.
Custard powders	Wheat, potato and rice flour, lead chromate, turmeric to enhance the yellow colour.
Coffee	Chicory, roasted wheat, rye and potato flour, roasted beans, acorns *etc.*, and burnt sugar (black jack) as a darkener.
Tea	Used tea leaves, dried leaves of other plants, starch, sand, china clay, French chalk, Plumbago, gum, indigo, Prussian blue for black tea, turmeric, Chinese yellow, copper salts for green tea.
Cocoa and chocolate	Arrowroot, wheat, Indian corn, sago, potato, tapioca flour, chicory Venetian red, red ochre, iron compounds.
Cayenne pepper	Ground rice, mustard seed husks, sawdust, salt, red lead, vermilion, Venetian red, turmeric.
Pickles	Copper salts for greening.
Gin	Water, cayenne, cassia, cinnamon, sugar, alum, salt of tartar (potassium tartrate).
Porter and stout	Water, brown sugar, *Cocculus indicus*, copperas, salt, capsicum, ginger, wormwood, coriander and caraway seeds, liquorice, honey, Nux vomica, cream of tartar, hartshorn shavings, treacle.

be proven. It is also worth noting that fraud can include the placing of food which is microbiologically unfit on to the market. Such cases are not listed in RASFF under compositional standards.

Methods of detection have changed since 1860, and consequently, new areas of adulteration are being found, particularly in meats, prepared foods and snacks, and fats and oils. More recent large-scale fraud, as noted in Sudan dye[6] in chilli powder (2005), chicken[7] (1990s–2008), melamine[8] in Chinese baby milk (2008) and Horsegate (2013), have all demonstrated the world-wide complexity of supply and the potential for fraud on a major scale. Food fraud is thought to affect up to 10% of all the foods we eat in the developed world and up to 20% in the developing world. The cost of food fraud globally has been estimated to be £28 billion each year.

During Horsegate, several leading figures in the retail industry suggested that cheap food was the reason why the fraud occurred.[9] This almost implies that traders simply responded to market demand and blame was laid, in particular, at the door of local authorities given their relentless drive towards ever cheaper school meals. This, in turn, made it more likely that meat would be substituted. Some argue that lower costs lead to the need to reduce overheads and this, in turn, leads to less confidence that food is as described.

A recent BBC documentary, 'Victorian Bakers',[10] suggested that the adulteration of bread may simply have been in response to customer demands for cheap food. Without adulterants, the food would simply not have been available, leading to malnutrition and starvation. Unsurprisingly, the enforcement members of the research team did not subscribe to these arguments. I am not confident about the Victorians as I doubt that they sought sub-standard food. Indeed, Accum's book was a best-seller and was popular because it enabled consumers to identify fraud, and those who were starving might have gladly eaten anything. I would be interested to watch any defence presented in court that Horsegate was in response to customer demand! The food was not of the *'nature, substance or quality so demanded by the consumer'* and therefore in breach of the 1990 Food Safety Act (and all previous Acts going back to 1860). The prosecution has now proven that there was intent to defraud passing off horse as beef. The defence never argued by way of mitigation that the consumer deserved this because they sought cheap food!

Game Theory, developed by John Von Neumann,[11] predicts that as we try to think like a food fraudster and change defence systems, this will alter the behaviour of fraudsters. Head of Fraud for the City of London Police, Commander Steve Head has said that crime continuously adapts and fraudsters

Table 7.2 Statistical analysis of the cost of fraud presented by Commander Steve Head.

£73bn	Cost of fraud to the economy	National Fraud Authority (NFA), Annual Fraud Indicator (AFI), 2013[12]
£9.9bn	Perpetrated by Organised Crime Groups	IO soc. strategy 2013[13]
24 800	Reports of crime	National Fraud Intelligence Bureau (NFIB) 2014/15
7%	Increase	(NFIB) 2014/15
40–70%	Cyber-enabled	(NFIB) 2014/15
80%	Cyber-crime not reported to police	(NFIB) 2014/15

are particularly good at adaptation. 70% of fraud is now cyber-enabled, and yet despite the threat, vulnerability is growing. The resources needed to fight fraud are diminishing and so the focus must be on prevention strategies. At a recent seminar at the University of Portsmouth, Steve Head presented the statistics on fraud shown in Table 7.2.

Fraudsters will use their knowledge to change strategy and operating techniques based on their information and perceptions. Accordingly, those working on fraud defence must do likewise. Unfortunately, the two will never be equal, and this asymmetry of information will always require defence systems to continuously adapt.

In summary, fraud is not new; it will remain and is likely to increase. In true game theory style, the next section will deal with the fraudster's opportunities to attack, and the third will look at defence strategies to ensure the food industry will be less likely to be a target and so could be seen as the winners.

7.3 ATTACK

Research published by PKF Littlejohn states that only 3% of fraud is detected[14] and that it focusses on the high volume, low value areas because they are harder to detect, *e.g.* horsemeat substitution for beef. Chris Elliott and Sara Steed[15] state that food fraud affects around 10% of the foods we eat and costs around $40 billion each year.

7.3.1 The People Involved

There are two key types of fraudster which the research group focussed upon:

- Opportunistic fraudsters; and,
- Organised fraudsters (criminals who commit more organised and larger food fraud).

The National Crime Agency (NCA) defines organised fraudsters[16] as follows:

> 'Organised crime can be defined as serious crime planned, coordinated and conducted by people working together on a continuing basis. Their motivation is often, but not always, financial gain. Organised crime group structures vary. Successful organised crime groups often consist of a durable core of key individuals. Around them is a cluster of subordinates, specialists, and other more transient members, plus an extended network of associates. Many groups are often loose networks of criminals that come together for a specific criminal activity, acting in different roles depending on their skills and expertise. Collaboration is reinforced by shared experiences (such as prison), or recommendation from trusted individuals. Others are bonded by family or ethnic ties – some "crime families" are precisely that.'

Both opportunist and organised food fraudsters will seek to repeat the fraud and have contacts and links to the industry they defraud as well as knowledge of processes. They try to assess the likelihood of detection and determine the best opportunities, based on their 'limited' knowledge. It should be noted that the information they have may not be accurate or up to date. Game Theorists refer to this as asymmetric information; fraudsters will not have the latest, most up-to-date information regarding the work and capabilities of detectors and the detectors will not know where or when fraudsters might strike.

Opportunists are defined as those less likely to repeat the fraud and may exist within the food chain. They are likely to offer less of a risk and be controlled by in-house monitoring.

Opportunists can have a devastating effect/impact, but the overall impact will be very much greater from an organised fraudster who will have the ability to repeat the crime, often on a large scale.

Opportunistic fraudsters may strike anywhere throughout the supply process. Hirschauer and Musshoff[17] identified that some German farmers were more likely to use 'illegal chemicals' on their crop at harvest time if they thought their crop was being amalgamated with others before testing, as they thought this would render it more difficult for enforcement to detect their transgression. This relied on asymmetric information, as they could not be sure that the product would be amalgamated or that it wouldn't be tested.

The more complex the food fraud, *e.g.* chicken fraud,[3] the more likely that it would be undertaken by more informed/organised fraudsters, especially where the profitability is higher.

7.3.2 The Foods Involved

All foods can be the subject of fraud.

7.3.2.1 When is Food Most Vulnerable?. Building on the data from previous food fraud, Figure 7.1 shows the key weaknesses in the food processing chain.

The model (Figure 7.1) was developed by Naughton and colleagues[18] and shows key transgressor nations and key detector nations. The depth of colour indicates increased frequency of transgression or detection. This network analysis tool was further developed by Petroczi *et al.*[19] to show trends of transgression by nations and provide a history of non-compliance of supplier and country of supply, both key factors for food fraud. Previous issues are replicated or developed; for example, spices have been adulterated by the addition of substances to improve appearance for over 150 years. It is worth noting that the RASFF[20] data used in the model shown in Figure 7.1 are focussed on risk to human health, so frauds which do not offer a threat to health may not be recorded in this system. A fraud system is being developed (Table 7.3).

The two key areas of risk are processing after harvest and manufacture. Researchers agreed that the risks of fraud increase

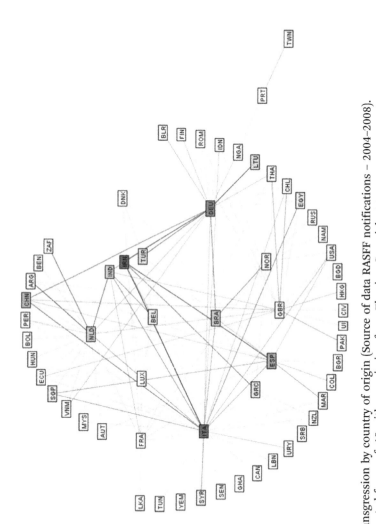

Figure 7.1 Transgression by country of origin (Source of data RASFF notifications – 2004–2008). Adapted from ref. 19 with permission from Elsevier, Copyright 2010.

Table 7.3 Stages and opportunities for fraudsters.

How and where		When (stages – from farm to fork)									
Commodities	Types of fraud	Growing/ harvesting	Distribution after harvest	Processing e.g. milling	Distribution	Packaging/ processing e.g. combination	Distribution	Wholesaler	Distribution	Point of sale	Waste
Additives: Sweeteners, salt, yeast, gelatine, ascorbic acid, acidity regulators. *etc.*	Substitution and dilution			X[a]		X					
Honey, fruit juices	Adding sugars			X		X					
Alcoholic beverages, coffee, tea and cocoa, confectionary, dietetic foods, fats and oil, feed, fruits and veg, fruit juices, milk, honey, herbs and spices, prepared dishes, soups and broths	Physical, biological or chemical treatment			X		X					
Alcoholic beverages, dietetic foods, fats and oil, feed, fruit juices, herbs and spices, milk, honey, meat (mince), prepared dishes, soups and broths	Dilution/mixing	X		X		X					
Meat, fish, oil	Disguise species			X		X					
Honey, saffron, spelt flour	Disguise flora	X		X		X					
Meat, fish, spices	Disguise spoiling e.g. irradiation	X		X		X					
All commodities	Mislabel	X	X	X	X	X	X	X	X	X	X

[a] Manufacture.

earlier in the processing chain. However, this is not always the case. Both FSAs in Ireland and England commissioned a food chain vulnerability assessment for the white fish supply chain and noted that the risks of substitution with other species were not only at the processing stage but also at the wholesaler. A further key factor for fraud is how the food is processed. The more food is processed, thus making the fraudulent activity harder to detect, the more likely fraud will occur, for example fish for use in processed meals.

For processed foods, this can be further analysed for risk by sectors (Figure 7.2).

7.3.2.2 Additional Key Factors Affecting the Likelihood of Food Fraud.
1. Supply and demand:
    - commodity prices, especially where there may be substitution of one species for another or where country of origin, or product of designated origin, is attracting a premium;
    - crop failures that lead to increasing raw materials costs, especially where these are not reflected by changing prices in the supply chain;
    - fishing restrictions or pressure on specific catch methods leading to fish shortages; and,
    - rapidly increasing demand. Research by the main honey producers' organisation in New Zealand, from where almost all the world's manuka honey comes, revealed staggering variations in the volumes produced and the volumes sold as 'manuka'.
2. Changes in legislation.
3. Relationship with suppliers:
    - good close-working relationships will result in trust which may be abused when circumstances change. Equally, lack of a good working relationship may promote fraud – NB. strong link with longer, more complex supply chains;
    - longer, more complex web/distribution chains which may include brokers and subcontracting of supply; and
    - the history (and any changes) of the supplier from a criminal records perspective.

Thinking Like a Food Fraudster

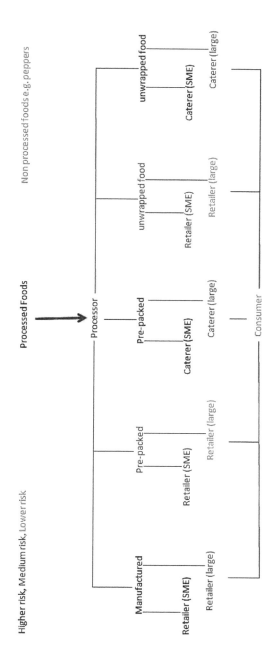

Figure 7.2 Risk analysis of fraud in processed foods.

4. Capability/likelihood of detection:
   - high costs of detection (analysis) or a poor reliability of the testing regime (*i.e.* foods that are simply assessed for one parameter such as nitrogen content or colour); and,
   - the perceived likelihood of being checked.
5. Profitability of the fraud:
   - availability of ingredients/products/food, ease and cost of substitutes;
   - the availability of waste products which can be used to mask quality assessments, *e.g.* melamine (baby milk);
   - value of the food: high volume, high value, high differential; and,
   - price premium, for example, products of designated origin or health benefits.
6. Risk and perception of risk:
   - perceived likelihood of detection;
   - penalties for being caught; and,
   - try once, perhaps on a relatively small scale and then repeat (note RASFF data).

Fraud can occur anywhere, anytime, and the complexity of food makes it harder to detect. Fraudsters are intelligent and informed and are likely to be closely involved in the process and seek a profit.

7.4 DEFENCE STRATEGIES

7.4.1 The Importance of Communication: A 'Burglar Alarm on Britain'

The UK would benefit from creating the perception of a 'burglar alarm on Britain' to demonstrate the UK's resilience against the threat of food fraud. This would apply equally to any other member state. The theory is that alarms are used as an effective deterrent against crime. Perhaps the key part of an alarm is the box demonstrating from the outside that a property is protected. Therefore, a member state would communicate that it will not tolerate food fraud and food companies must do the same to all involved in their supply web by actively monitoring to stop it and share the information so that prosecutions will be the result whenever fraud is identified. It is vital, therefore, that the food industry develops and maintains a good communication strategy so that, when fraud is detected, a proactive and timely media strategy highlights that the fraud has been thoroughly investigated and strong action taken with stringent penalties imposed.

Government must also adopt the same stance by taking a lead and actively adopting strategies such as those set out in Professor Elliott's report and communicating 'successes' to support the food industry and protect the public. And, finally, to be at its most effective, the burglar alarm principle then needs to be incorporated across the EU given the pan-European nature of the food chain.

7.4.2 A Vital Role for Government (Member States)

It was agreed that, to maximise effectiveness, each member state needs to implement the following:

- better communication systems to announce new information to ensure that perception and likelihood of detection are high;
- change legislation to enable more stringent penalties for food fraud and link penalties to the value of the fraud, including the ability to seize assets gained through criminal activity using the Proceeds of Crime Act (POCA);
- fund a specialist enforcement team(s) capable of pursuing fraud cases, *i.e.* able to prove intent and with powers which are not available under current food safety law, such as use of covert surveillance techniques, *e.g.* tracker devices on vehicles, seizure of telephone records or bank account details, CCTV monitoring *etc.* – in the UK this will be the FSA's National Food Crime Unit;
- ensure that police, customs and others, particularly those with expertise in cyber-crime, share intelligence and resources and work together with food enforcement officers;
- establish better traceability/tracking such as provenance systems.[21] Provenance is defined as the chronology of ownership, and IT-based systems can provide data on the source of ingredients, their travel routes and times *etc.*;
- harmonise world-wide laws to reduce the opportunities to use locally compliant food which fails to meet more

stringent EU safety standards; for example, prawns farmed in India meet local standards if they contain antibiotics (chloramphenicol) but fail to meet EU safety laws, which ban the use of antibiotics in food;
- register food brokers as food businesses so they can be audited by enforcement agencies;
- establish intelligence systems which help industry understand the newly available tests which are being developed – for example, metabolomics or time-of-flight mass spec fingerprinting for food authenticity – and how to interpret what their limits of detection are utilising EU reference laboratories to train both industry and enforcement;
- establish collective intelligence systems which will assess news releases, weather, supply and demand data, tax data, customs data and police intelligence, and ensure that all databases used by enforcement facilitate easy searching and network analysis;
- set up systems for international co-operation;
- encourage food businesses to set up anti-fraud cultures and share information with enforcement agencies;
- host whistle-blowing systems to ensure that people reporting issues in their industry can do so anonymously to an independent body;
- ensure that audits which are used to protect against fraud do not rely on form filling (tick box) but focus on discussion using open questioning which will reveal more about processes *etc.*; and,
- help food fraud enforcement to develop systems which 'follow the money' to trace the fraudsters who benefit.

7.4.3 A Key Role for Industry

There are many useful tools which are available to help the food industry protect itself against fraud. These include VACCP,[22] Carver and Shock,[23] Food defence plan builder,[24] BRC,[23] PAS 96[25] and FDF five steps.[26] One caveat though; I haven't applied all of these tools. Food industry friends have informed me that the following approaches are vital:

- Shorter food chains: in response to Horsegate, some food businesses have actively reduced their food supply webs and formed closer relationships with a small number of local suppliers with the intention that fraudsters could not easily infiltrate these relationships and, as a result, monitoring, auditing and testing were easier, as was sharing results and intelligence. In addition, the larger contracts developed were too valuable to the supplier to risk losing as a result of fraud. The organisation I worked for implemented a similar strategy for the supply of school meal food. This system worked well when a world-wide or out-of-season supply is not required by customers. To replicate such a system across the world on a macro-scale is labour-intensive and costly, but still necessary.

- Exert pressure on fraudsters: all food manufacturers can do is put as much pressure on the fraudsters as possible by identifying those materials which are 'high risk' and then implementing reasonable mitigating controls, such as close and regular auditing, restricted supply chains, comprehensive testing schedules *etc.* A single business can only have a negligible impact over a fraudster operating across the world. They can put the 'burglar alarm' on their 'own house' and make the fraudsters look elsewhere; yet, if a whole industry or sector work together, they can achieve so much more.
- Seek help from FSAs: food businesses do not have sufficient expertise in predicting either the impact of a poor harvest in one region, or supply and demand predictions from a fraud perspective. It has to be accepted that food businesses can't be expected to be experts in every area of the world-wide supply chain. What is needed is a simple source of (interpreted) information which can be utilised by UK/EU manufacturers, *e.g.* an Intelligence Hub.
- Seek scientific help with analytical testing: as an industry, huge amounts of time and financial resources are now being focussed at testing raw materials for 'potential' contaminants. This equates to trying to find an unknown needle in an unknown haystack! It will be very easy for the entire food industry to focus all its testing resources in one area and completely miss another. What is needed is a portal for communicating results across the industry so that the results from one food business operator can be

used by another as part of their due diligence and *vice versa*. An industry-coordinated approach is required, with clear direction from the FSA, in order to prevent a complete 'scattergun' approach and to focus resources and share outcomes from testing regimes.

7.4.4 The Weakest Link

The weakest link will be smaller food businesses, including caterers who don't have the resources they need to defend against fraud. They need help from associations and networks and, in particular, the government. Enforcement agencies should see these companies as higher risk and commit resources to monitor the possibilities of fraudulent activity in these sectors.

The threat from food fraud will be ever present. No business can work in isolation to reduce the threat and neither can government. It is essential that all work in concert, sharing intelligence, expertise, information and new developments. The only option is to demonstrate that fraud will not be tolerated and that all involved are strenuously working towards identifying fraud and prosecuting those involved, utilising the latest techniques, in the hope that fraudsters will get the message that they will be caught and punished.

There are definite signs of sound progress in the UK; to date, these include:

- establishment of the National Food Crime Unit within the FSA;
- more prosecutions in the Crown Court rather than simple enforcement action or service of notices where adulteration or substitution occurs;
- more serious criminal action, such as fraud charges, thereby attracting longer sentences, such as imprisonment or significant financial penalties;
- restraint and seizure of assets gained through criminal activity; and
- increased media attention in both specialist areas, but also raising public awareness and reaction to demand action and prevent the likes of Horsegate happening again.

REFERENCES

1. Food Fraud Reference Sheet, http://foodfraud.msu.edu/wp-content/uploads/2014/08/flyer-FF-Reference-Sheet-Final.pdf [accessed August 2018].
2. Food Safety Act 1990, http://www.legislation.gov.uk/ukpga/1990/16/contents [accessed August 2018].
3. G. Taylor, *Forensic Enforcement: The Role of the Public Analyst*, The Royal Society of Chemistry, Cambridge UK, 2010.
4. F. Accum, A Treatise on Adulterations of Food and Culinary Poisons - The Original Classic Edition, Tebbo, 2012.
5. RASFF – Food and Feed Safety Alerts, https://ec.europa.eu/food/safety/rasff_en [accessed August 2018].
6. Food dye scare sparks largest recall in history, https://www.telegraph.co.uk/news/uknews/1484071/Food-dye-scare-sparks-largest-recall-in-history.html [accessed August 2018].
7. Food Law News – UK – 2009, http://www.reading.ac.uk/foodlaw/news/uk-09014.htm [accessed August 2018].
8. Timeline: China Milk Scandal, http://news.bbc.co.uk/1/hi/7720404.stm [accessed August 2018].
9. I don't eat value food because it doesn't contain much meat': Boss of budget supermarket Iceland in shock claim as Waitrose chief warns of cheap food risks, http://www.dailymail.co.uk/news/article-2280005/Horsemeat-Boss-budget-supermarket-Iceland-shock-claim-Waitrose-chief-warns-cheap-food-risks.html [accessed August 2018].
10. BBC Two; Victorian Bakers TV Program, http://www.bbc.co.uk/programmes/b06vn7sj [accessed August 2018].
11. John von Neumann, https://en.wikipedia.org/wiki/John_von_Neumann [accessed August 2018].
12. National Fraud Authority, Annual Fraud Indicator June 2013, https://www.gov.uk/government/uploads/system/uploads/attachment_data/file/206552/nfa-annual-fraud-indicator-2013.pdf [accessed August 2018].
13. Serious and Organised Crime Strategy 2013, https://www.gov.uk/government/uploads/system/uploads/attachment_data/file/248645/Serious_and_Organised_Crime_Strategy.pdf [accessed August 2018].
14. PKF LITTLEJOHN LAUNCHES A NEW CHARITY COUNTER FRAUD GUIDE, https://pkfcounterfraud.wordpress.com/ [accessed August 2018].

15. Addressing Complex and Critical Food Integrity Issues Using the Latest Analytical Technologies, http://www.waters.com/waters/library.htm?lid=134877763&locale=en_GB [accessed August 2018].
16. Organised crime groups, http://www.nationalcrimeagency.gov.uk/crime-threats/organised-crime-groups [accessed August 2018].
17. N. Hirschauer and O. Musshoff, *Food Policy*, 2007, **32**(2), 246–265.
18. T. Nepusz, A. Petróczi and D. P. Naughton, *PLoS One*, 2009, **4**(8), e6680.
19. A. Petróczi, G. Taylor, T. Nepusz and D. P. Naughton, *Food Chem. Toxicol.*, 2010, **48**(7), 1957–1964.
20. RASFF Portal, https://ec.europa.eu/food/safety/rasff/portal_en [accessed August 2018].
21. L. Moreau and P. Groth, *Provenance: An Introduction to PROV*, Morgan & Claypool Publishers, 2013.
22. GFSI Direction on Food Fraud and Vulnerability Assessment (VACCP), http://foodfraud.msu.edu/tag/taccp/ [accessed August 2018].
23. BRC Global Standard for Food Safety Issue 7 – vulnerability assessments, http://www.adas.uk/Service/brc-global-standard-for-food-safety-issue-7-vulnerability-assessments [accessed September 2018].
24. Food Defense Plan Builder, https://www.fda.gov/Food/FoodDefense/ToolsEducationalMaterials/ucm349888.htm [accessed August 2018].
25. Guide to protecting and defending food and drink from deliberate attack, https://www.food.gov.uk/sites/default/files/media/document/pas962017.pdf [accessed September 2018].
26. Food Authenticity: Five steps to help protect your business from food fraud, http://www.fdf.org.uk/food-authenticity.aspx [accessed August 2018].
27. The fight against food adulteration, https://eic.rsc.org/feature/the-fight-against-food-adulteration/2020253.article [accessed August 2018].

CHAPTER 8

The Bank Job

8.1 WILL THE MAJOR FOOD BUSINESSES AND REGULATORS LEARN FROM THE BANKING CRISIS?

A decade ago, the bankers 'blew up the British economy' and, through the multi-billion pound bailout courtesy of you and me, the public are still paying the cost. Discovering what happened in Iain Martin's brilliant book, *Making it Happen*,[1] made me appreciate that there are several parallels with other business sectors and some surprising similarities with food. Food businesses do not have quite the same impact on our economy as the banks in terms of pensions and financial stability, so should one or more collapse, it is highly unlikely it would be rescued by the government in the same manner. That should be a warning to the food industry. And, indeed, insurance companies and investors might, as we have seen elsewhere, be quick to pull the plug on their investment to cut their potential losses and the risk to their reputation, by association. The banks are, and were, different. There are valuable lessons for all sectors, and food businesses in particular, to learn from the 2008 banking crisis and, more importantly, how it occurred.

It is hard to predict precisely where the banking crisis started. Whilst the first principle of business is that investors seek to maximise their wealth, it should not be at all costs. In the latter part of the last century, many bank executives had an insatiable appetite for increased profitability and bonuses. Fred 'the shred' Goodwin, a nickname he apparently liked, has been credited as the major protagonist, although to be fair to him, there were many others. He drove the Royal Bank of Scotland (RBS) towards the 'big league' by competing with others across the globe, striving to build the biggest bank in the world. It started in 1998 when he became deputy chief executive of RBS and set about revolutionising the bank by focussing on dramatic increases in profit and aggressive expansion, primarily through acquisition and hostile takeovers of other global banks, some suggest at all costs. Growth was the only target. Consequently, the bank took on unsustainable debts from other organisations – for example, sub-prime mortgages – leading to unsustainable debt followed by crippling costs which ultimately led to a catastrophic collapse. Goodwin was obsessed with the wrong kind of detail, failing to focus on the bigger issues within the company, a bully to his staff and driven by an egotistic urge to trample his enemies. He enjoyed cosy chats with another Scot, the Chancellor Gordon Brown.[2] Fred fell from grace and, having been knighted the golden boy of business, became the pariah of the decade. What he left was a very valuable lesson for others to learn from, but will they?

In March 2000, Don Cruickshank's[3] report on competition in banking suggested that banks should not be allowed to grow too large. The report was ignored by the UK government who felt that the industry was over-burdened with red tape and regulation. The golden age of banking continued with new-found freedom from burdens of bureaucracy. In June 2007, at his final Mansion House speech, the Chancellor Gordon Brown[4] exulted in reduced bureaucracy and freedom for the banks. Does that sound familiar? There have been many calls from food industry managers and representatives lobbying government to reduce red tape and regulation and that they should be free to self-regulate and demonstrate compliance through invited third-party assurance. Only a few months after the Mansion House speech extolling the virtues of business free from shackles of regulation, holes in RBS finance were becoming a chasm as they strived to take over ABN AMRO. This pushed RBS over the financial precipice towards oblivion, precipitating a house-of-cards-style collapse across the whole financial system. A world-wide collapse became evident. The Chancellor and his team, however, emerged well from a rescue which helped to prevent a complete collapse of the whole financial system, albeit at a massive cost to the public. It has been said that the deregulation agenda, coupled with self-policing, led to a 'spiv culture' which, contrary to the perceived wisdom, did not encourage businesses to self-regulate. After all, spivs are not renowned for self-regulation; at least not a regulation that is in the interest of their customers.

At that time, the financial world believed 'big is beautiful'. Efficiency, effectiveness and a complexity of supply were necessary adjuncts to achieve this rapid growth and profitability. To help businesses, the government focussed more on risk-based enforcement and earned recognition, *i.e.* leave the companies to manage themselves as they can be trusted to operate in the best interest of consumers. The government of the day created a division of responsibilities between government agencies, leading to a lack of clarity for both regulators and enforcers [Financial Services Authority (FSA) and the Bank of England] which, in turn, meant control of the banks became almost impossible. Some banks continued the push for more freedom by suggesting that modern banking was too complex for regulators and enforcers to fully understand and, therefore, banks should be granted more freedom. By their actions, the government of the day were supportive of these arguments.

Those of us in food enforcement at the time might have seen some parallels. Many food businesses focussed on growth and maximising profitability; in itself, not a problem. They called for deregulation and earned recognition, and I even heard some say that they have so many product lines that it is difficult for the enforcers and regulators to check and understand the food industry. In addition, some food businesses improved profitability through an increasingly complex food supply web. In response, the government suggested earned recognition, risk-based enforcement and deregulation as the way forward. This led to changing the roles of DEFRA, the Department of Health and the Food Standards Agency ('the other' FSA), which once again precipitated a division of responsibilities between local authority enforcement and national agencies, making it more difficult to achieve effective enforcement. As a result of austerity and possibly some confusion, I suspect all food authorities in the UK have reduced their focus (resources) on food enforcement – this continues as local authorities further cut budgets. Despite having the power to order a 'writ of mandamus', which allows the FSA to take action against food authorities who are not fulfilling their role as an enforcement body, to date no action has been taken. To be fair to the FSA, it is hard to clarify what insufficient action by a food authority means as the role is not sufficiently prescribed, but this only serves to increase the confusion plus, of course, the FSA hardly has the resources to take over the responsibilities of a food authority, and if it did step in, it would be seen as moving away from a government keen on deregulation, and the FSA has said on many occasions it has no intention of providing enforcement; it prefers to leave that to the local authority.

There is, however, key learning from the banking crisis; namely, food businesses:

- should not focus solely on profitability;
- should be careful where they cut costs, especially if these may impact on their ability to self-regulate;
- should not rely on food authority monitoring and intelligence as it diminishes; and,
- perhaps above all else, should reduce complexity of supply as much as possible so that it is easier to monitor.

Also, lessons for the government and regulators:

- if earned recognition is to be adopted then enforcement will need to be able to scrutinise systems, have access to confidential reports by the accrediting body concerning the inspection and set or agree the standards the accreditation body uses to inspect the food business – similar to those used by OFSTED for schools;
- regulators will need the power to insist that monitoring results are displayed. Currently, it is not mandatory to display the results of the successful Food Hygiene Rating System (FHRS), and therefore food businesses can choose whether

or not to display the results, yet it provides vital information to the consumer in an instant when they visit the business; and,
- government will need to ensure resources are available for enforcement to monitor the food businesses that do not embrace self-regulation/accreditation principles.

The year before the horsemeat scandal became big news, some Welsh food authorities did look for substitution of meat with horse and none was detected. None of the English or Scottish food authorities undertook this test. Was this due to the additional cost; LAs pay for each test carried out on food, so testing for horse in addition to sheep or goat would add to the overall sampling and analysis bill and push tight budgets even further? Or was it a lack of knowledge/intelligence? As we know, the Irish authorities did look and find horse in place of beef. The food industry itself was not focussing on this issue at the time as no results from their self-monitoring came to light. Clearly, a lot has been learned following the horsemeat crisis, especially following Professor Elliott's investigations and report, and the food industry is embracing the changes outlined within the report. The food industry needs to be aware that there is an increasing move towards criminal prosecution, where the sanctions are much bigger, rather than proceedings through Magistrate Courts. The public are demanding action and will expect penalties. I suspect culprits will not fall from grace and disappear in future but suffer criminal penalties.

If a food business fails, I don't think the government would shore them up in the same way as the banks. So, whilst food businesses currently work with regulators, enforcers and others, they need to remember the lessons from the banking industry, especially if tempted to cut their own self-regulation systems, and the regulators also need to heed the lessons learned.

REFERENCES

1. I. Martin, Making it Happen – Fred Goodwin, *RBS and the Men Who Blew up the British Economy*, Simon & Schuster UK, 2013.

2. Making It Happen by Iain Martin – Review, https://www.spectator.co.uk/2013/10/making-it-happen-by-iain-martin-review/ [accessed August 2018].
3. The Cruickshank report: Competition in the banking industry under scrutiny, https://uk.practicallaw.thomsonreuters.com/7-101-1508?service=competition&__lrTS=20180227095803064&transitionType=Default&contextData = (sc.Default)&firstPage = true&bhcp = 1 [accessed August 2018].
4. Speech by the Chancellor of the Exchequer, the Rt Hon Gordon Brown MP, to Mansion House, http://webarchive.nationalarchives.gov.uk/ + tf_/http://www.hm-treasury.gov.uk/2014.htm [accessed August 2018].

CHAPTER 9

Someone to Watch Over Me

Being watched can be both positive and negative. Watching a children's film with one of my grandchildren recently, I was reminded how often a pair of eyes are used to scare, be they used to good effect in a dark forest or in a tunnel and occasionally as a threat; I'm watching you! We see eyes as watching over us to ensure our safety, or watching us in a menacing or frightening manner perhaps to do us harm. Could the two 'i' of insurance and investors work in a similar manner, both as a help and a threat? Both bodies seek a trouble-free association and a profitable outcome. Insurers want to reduce their risk of exposure by ensuring their client takes the appropriate strategies to reduce the likelihood of a problem, and where the insurer doubts the capability of the client, they may decline to insure. I have been on the receiving end of questionnaires from insurance companies seeking information about our management systems and capabilities on health and safety risks, asbestos within our buildings, for example. They were quite rightly keen to review our systems and ensure they weren't likely to be exposed to additional unforeseen or unnecessary risks and claims.

The Lloyds Register Foundation announced in February 2018 that it is funding a foresight review on food safety. The Lloyds Register says it is one of the world's leading providers of professional services of engineering and technology. It has around

---

The Horse Who Came to Dinner: The First Criminal Case of Food Fraud
By Glenn Taylor
© Glenn Taylor 2019
Published by the Royal Society of Chemistry, www.rsc.org

33 000 clients, and the review will take a global perspective and make recommendations on research, innovation and the education policy and practice needed to plan for a healthy and safer future for all. I think the insurers are watching this area, and rightly so. Investors also monitor the organisations in which they invest, not just to assess whether the company is likely to continue to thrive but also to see if their management capability is sufficient to develop the company and continue to do so. Many large investors also have their own reputation to consider and would not want a 'problem child' to damage their business standing or their name. Therefore, both organisations can be seen as someone who may protect or close a company. This could be a powerful tool in the armoury of regulators; both investors and insurers will put pressure on companies to meet the required regulatory standards. Therefore, if either of the two regard a food business as too much of a risk, sometimes referred to as a 'toxic brand', they might remove their support, which might lead to the closure, sale or rebranding of the business.

This feels like a new development. Having had a life time in this arena, I believe that many of us in regulation and enforcement were surprised at how effective this strategy might be; certainly those I spoke with at the time echoed this opinion. The regulator can ask questions of a food business and, if they are less than happy about its response to an issue, release it as public information. This might then lead to an increase in the toxicity of the brand with its investors, insurers and consumers and is a form of naming and shaming which, although isn't new – far from it as legislation was enacted in 1872 to allow this – now has a key difference in the way in which we receive our news and information, almost instantaneously and from a plethora of uncontrolled sources.

Food companies and the regulator will both need to be very media savvy and be seen to react quickly to save their respective reputations. The regulator demonstrated this during the 'birdtable' meetings introduced towards the beginning of the horsemeat scandal, and likewise, food businesses who worked hard to communicate with the public as to how they were responding to the crisis. Only the owners and investors will know if this new approach by the regulator had an impact on one of the brands caught up in horsemeat; let's look at what happened to that company during the scandal.

Findus were one of the larger brands to get caught by the horsemeat fraud. They were far from the only brand or food business. Nevertheless, information was presented in the public arena by the regulator through Ministers. Owen Paterson, the Environment Secretary, made the following statement at the beginning of the horsemeat crisis (7 February), having taken the unusual step of naming/identifying Findus beef lasagne products.

> 'It is completely unacceptable that a product which says it's beef lasagne turns out to be mainly horsemeat. Consumers have a right to expect that food is exactly what it says on the label.
>
> The presence of unauthorised ingredients cannot be tolerated. This is especially true when those ingredients are likely to be unacceptable to consumers, or where there is any conceivable risk to human health.
>
> The responsibility for the safety and authenticity of food lies with those who produce it, and who sell or provide it to the final consumer. I know that food producers, retailers and caterers are as concerned as we are at the course of recent events.
>
> The FSA is urgently investigating individual suspicious incidents, which they have taken up with authorities and police across Europe.
>
> The FSA and Defra are also conducting a survey of processed beef products in the UK – including supplies to schools and hospitals – in order to assess whether there are any significant levels of improperly described meat.
>
> In addition, the food industry agreed on Monday at their meeting with Food and Farming Minister David Heath and the FSA, that they would share the results of their own testing with the FSA and make the results publicly available. David Heath is meeting major food businesses again for a further update on Wednesday 13 February.
>
> The FSA has today requested that producers and retailers test all their processed beef products by the end of next week for the presence of horsemeat, and for residues of the veterinary medicine "bute".
>
> The Food Standards Agency, Defra, and the Department of Health are working closely with businesses and trade bodies along the whole food chain to root out any illegal activity and enforce food safety and authenticity regulations. Consumers

can be confident that we will take whatever action we consider necessary if we discover evidence of criminality or negligence.'[1]

I was surprised when this happened. It seemed that Findus were the only company named and placed in the headlines, and yet many others were caught by the fraud too. Certainly, the media named many companies;[2] for example, 19 major brands were 'named and shamed' by the BBC. On the same day, Findus were publicly rebuked over horsemeat. *'"The meat content of some beef lasagne products recalled by Findus was up to 100% horsemeat", the Food Standards Agency has said.*'[3] Two days later the following headline could be seen in the Daily Mirror: 'Why the delay? Findus left horse meat lasagne on shelves for SEVEN days. Environment Secretary Owen Paterson was ordered back to London from the Midlands by David Cameron to take control of the fiasco'.[4]

Then, a week later, to compound things further, the BBC followed with questions like: 'Can brands like Findus recover?'.[5] At the same time, Findus were criticised for their management and response to the horsemeat crisis.[6] It seemed the death knell had sounded, and eventually, the *Metro* reported the demise of the name: 'Goodbye Findus crispy pancakes: Brand dogged by horsemeat scandal is to disappear'.[7] I can't help but feel sorry for Findus as they were far from the only company to be defrauded in the horsemeat scandal; it may look to some as if the FSA focussed on Findus early on, and did that result in Findus finishing? I guess the argument would be that the regulator is merely working in an arena of openness and transparency, as all in government are expected to do. Perhaps Findus could have responded better to the public to inform them of the measures the company were taking to avoid a recurrence. Certainly, other food businesses actively embarked on a public relations exercise. Tesco showed the way with an exemplar letter of apology, published in a full-page advert in all major newspapers. The advert read as follows:

> '**We Apologise**
> *You have probably read or heard that we have had a serious problem with three frozen beefburger products that we sell in our stores in the UK and Ireland. The Food Safety Authority*

of Ireland (FSAI) has told us that a number of products they have recently tested from one of our suppliers contained horsemeat.

While the FSAI have said that the products pose no risk to public health, we appreciate that, like us, our customers will find this absolutely unacceptable. The products in our stores were Tesco Everyday Value 8 x Frozen Beef Burgers (397g), Tesco 4 x Frozen Beef Quarter Pounders (454g) and a branded product, Flamehouse Frozen Chargrilled Quarter Pounders.

We have immediately withdrawn from sale all products from the supplier in question, from all our stores and online.

If you have any of these products at home, you can take them back to any of our stores at any time and get a full refund. You will not need a receipt and you can just bring back the packaging.

We and our supplier have let you down and we apologise.

If you have any concerns, you can find out how to contact us at the bottom of this page, or go to any of our customer service desks in-store, or ask to speak to your local Store Manager.

So here's our promise. We will find out exactly what happened and, when we do, we'll come back and tell you.

And we will work harder than ever with all our suppliers to make sure this never happens again.

TESCO'[8]

Not all food companies caught by a major issue have disappeared. Perrier, the bottled water company, looked like losing their fizz in 1990 when they withdrew their stock from supermarket shelves due to benzene contamination. This was found by North Carolina health officials and the levels were miniscule, but the public reacted, some might say irrationally, based on the press releases, despite the levels offering no significant risk to health. The benzene was a 'natural-occurring' contaminant and easily removed by filtration. The company responded well, withdrawing stocks for a number of months, and as a result, when they returned a few months later, they lost their place as the outright market leader in a market that they had almost single-handedly developed. Others now lead in the fizzy bottled water market, which was not shrinking. Perrier were sold to Nestle two years later;[9] nevertheless, Perrier survived an onslaught.

It seems that investors may decide to cut ties with a company if they feel a brand becomes too toxic. All food businesses need the loyalty of customers, and if there are concerns about a name, then a company will need to do something as profitability is likely to be affected. During the horsemeat crisis, many large household names must have been concerned about their market value and longevity, but as so often happens, most managed to survive the crisis, mostly by communicating their responses to the issue.

I wonder how this new tool in the armoury of the regulator will be used in future and how food businesses will react to the new strategy. If the regulator does use this strategy then industry will need to learn the power of responding in a timely way through the media to avoid adverse media coverage and speculation.

9.1 TSB: THE FRAUDSTER'S EYES ARE WATCHING YOU

I have a friend who runs a large property company. We were discussing fraud, and he is concerned as to how he would protect himself against it. He said that it is hard as you don't know where it will come from within your organisation. Yes, you can have systems against general fraud, but specific types are harder to manage, especially if you haven't encountered it before. How do you prepare for the unknown without spending huge sums of money? It's a bit like the childhood bogeyman; you have heard of it, you are frightened of it, but you cannot be sure you have all the bases covered. I guess that has been demonstrated by the recent IT banking issue at TSB.

The fraudsters were watching and have been so quick to respond to the recent issue with computer systems following an upgrade at TSB and the resultant confusion. There has been an alleged ten-fold increase in phishing emails.[10] Once fraudsters have banking details from a phishing email, they then move on to the next step. The BBC states that fraudsters have been updating customers' phone numbers with new ones that are not in the possession of the customer but belonging to the fraudster. Then, when the bank supplies a code *via* a text to the customer following a request for a transfer of money *via* the online banking system, it is the fraudster who receives the text codes.[11] The fraudster is then free to transfer funds out of the customer's account, in some cases totalling thousands of pounds. Allegedly, on one occasion, this was carried out whilst the owner of the account was watching online, whilst waiting to report the issue to the fraud team.

If customers are sensitive about food, I suspect they are even more nervous about their money and their bank accounts. Recently, the media has been quiet about TSB's problems so I assume the company must have resolved the issue. The stuff of nightmares for both customers and banks, and I would imagine a bank could quickly lose customer loyalty if things were not resolved quickly.

9.2 WHO WATCHES EU FOOD SAFETY?

Dr Tamas Nepusz and Professors Andrea Petroczi and Declan Naughton worked together to produce a series of papers in 2009 on network analysis using the RASFF notifications for issues relating to food safety.[12] I joined the team and analysed RASFF data, and we produced a paper in 2010[13] which concentrated on food notifications and whether it was possible to build a tool for enforcement and others to manage food data better. This next section is based on that published research.

The data provided by member states to the RASFF database are enormous, complex and rapidly evolving. During the period 2000–2007, the number of notifications increased from 800 to 7000 per annum, each notification containing a plethora of information. Analysing the relationships in these data is difficult to achieve and to give meaningful interpretations. The enormity of the data, together with rapidly emerging trends, necessitates a different approach to analysis of the data. The hope was, and still is, that a new system would enable rapid, real-time responses by health and enforcement professionals to avoid further illness or fraud and be a tool to help interested parties, be they food businesses regulators, or anyone with an interest in the issue, to be able to analyse the data succinctly. A network analysis tool does facilitate capturing the data, analysis of trends and examination of possible interventions and could, if the EU chose to invest in such a system, improve the management and response to the data (Figure 9.1).

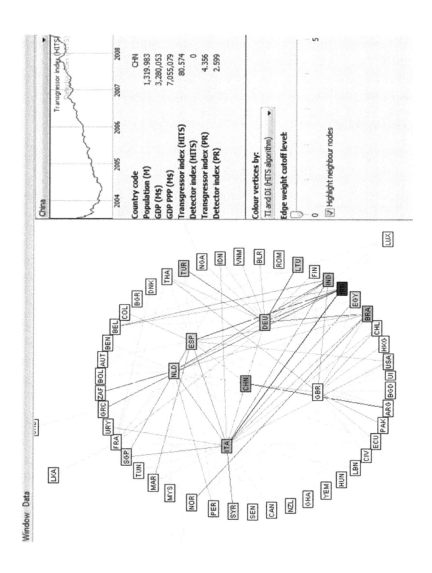

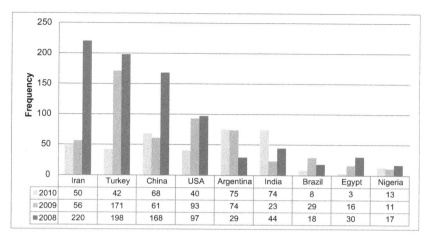

Figure 9.2 Number of mycotoxin notifications by country from 2008–2010. Reproduced from ref. 14 with permission from Wageningen Academic Publishers, Copyright 2011.

Figure 9.2 details an analysis of all mycotoxins, including aflatoxin. It covers the later time period of January 2008 to December 2010 and shows that the issue in Iran had become more controlled, with less notifications appearing on the database in subsequent years.

Figure 9.1 Network analysis of the data on the database from January 2003–August 2008. It shows detector indexes (DI) and transgressor indexes (TI) [two algorithms are used to calculate each: Google's Page Rank algorithm and Kleinberg's HITS algorithm. This shows increasing issues in the same manner as a Google search, with the total of all DIs and TIs normalised to 1 to help show the relative changes]. The strongest detectors and transgressors are shaded in the most; the more a country transgresses, the darker it becomes. By changing the relative weights, only the strongest relationships would be shown (darkest lines between actors). During the period chosen, there were issues with aflatoxin, a carcinogen from *Aspergillus* mould found in Iranian peanuts, which may have been poorly stored, and as a result, Iran is shown as a major transgressor (darkest shade). It can be seen that there are lines between Iran (IRA) and all the major detectors, which means that nuts from this source were imported through the ports of each of the detectors. There were 1938 notifications on the RASFF dataset relating to this issue during this period.
Adapted from ref. 13 with permission from Elsevier, Copyright 2010.

Our research[13] showed that four member states were the gatekeepers of EU food safety. They were Italy, Germany, UK and Spain (in order). A subsequent analysis of recent years has shown the order of detection to change from time to time, with Germany slipping into third place behind the UK (in the period 2010–2017). There appeared to be a relationship between those with strong frontier ports and their ability to detect transgression; unsurprising given that most of the issues on the RASFF database relate to food incoming into the EU from external providers and that some of the key frontier ports would provide food for neighbouring member states (food is free to travel throughout the EU once it has been accepted by a frontier port). However, it should be noted that both Italy and Germany are an exception to this rule as their monitoring on-the-market is stronger than most member states, suggesting that their focus on enforcement is stronger than most member states. The UK imports around 20% of its food from outside of the EU (Figure 9.3), but around 80% of the notifications relate to that food; the UK's monitoring on-the-market is very much weaker.

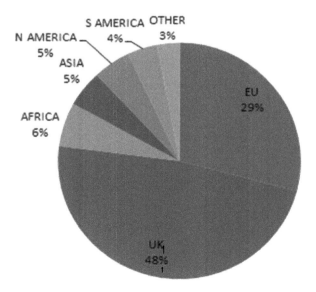

Figure 9.3 UK sources of imported food in 2006. Data from ref. 13.

The UK has always been a strong contributor to the RASFF database, particularly due to the strength of its port health monitoring, *i.e.* monitoring at the ports. A more detailed analysis of the UK's more recent performance has shown continued strength at the border inspection posts.

9.3 A COMPARISON OF THE TOP THREE GATEKEEPERS OF EU FOOD SAFETY FROM 2003–2017

9.3.1 Italy

The Italians have maintained pole position as the leading gatekeeper of EU food safety in terms of the numbers of RASFF notifications they raise each year. Their monitoring at the border is not identifying as many issues as it used to and their on-the-market monitoring remains very vibrant. Overall, the number of detections is falling, which may show less investment as austerity bites (Figure 9.4).

9.3.2 UK

The UK has moved into second place in the league of gatekeepers of EU food safety based on the numbers of RASFF notifications raised each year (Figure 9.5). Atypically, the total number of detections continues to increase. This is due to detections at the border inspection posts, the points of entry to the EU.

In recent years, the on-the-market monitoring carried out by environmental health and trading standards officers has resulted in a much lower number of notifications and the trend continues to fall (Figure 9.6). It is likely that this reflects the cuts in expenditure by local authorities.

9.3.3 Germany

The Germans have slipped to third place in the list of gatekeepers of EU food safety in terms of the numbers of RASFF notifications raised.

Companies are increasing their effectiveness in raising notifications, which is interesting when compared to the UK where there was a call by Elliott for companies to share their monitoring. I would have expected this to lead to more RASFF

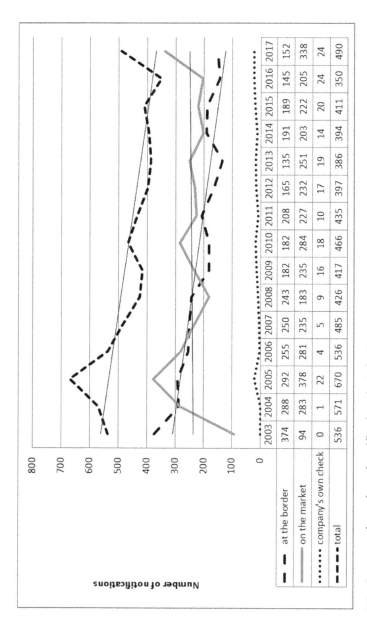

Figure 9.4 Source and number of notifications in Italy 2003–2017.

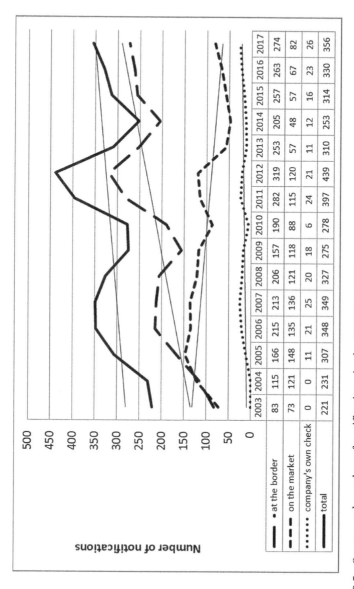

Figure 9.5 Source and number of notifications in the UK 2003–2017.

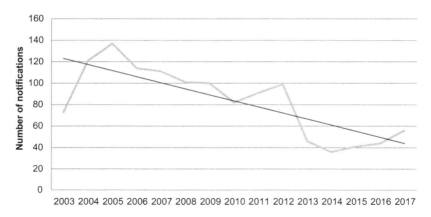

Figure 9.6 UK notifications arising from on-the-street monitoring 2003–2017.

notifications in this area. Overall, the numbers of RASFF notifications raised by Germany have fallen considerably as austerity bites (Figure 9.7).

9.4 SUMMARY

It seems that, in the leading member states, the 'someone to watch over us' is struggling to keep their eye on the ball as austerity bites. The UK has successfully continued with border inspection post monitoring despite the fact that, in recent years, funding for local authorities from the FSA has been withdrawn. Local authorities in the UK are clearly doing less monitoring on-the-market, and this is leading to fewer notifications. In the UK, it doesn't look as though industry has taken up the gauntlet thrown down by Chris Elliott, and possibly the only eyes watching over us are investors, insurers, media and port health. The recent announcement of more money for the fraud team will hopefully result in an increased threat to fraudsters, but this probably will not lead to more RASFF notifications. [NB. The horsemeat fraud and subsequent EU-wide monitoring led to 80 RASFF notifications for horse substitution.]

However, as we negotiate for Brexit, we must ensure that we have effective monitoring at the border, be it ours or across the EU, if there is free passage of goods into the UK from the EU. I strongly recommend the UK borders maintain their strength.

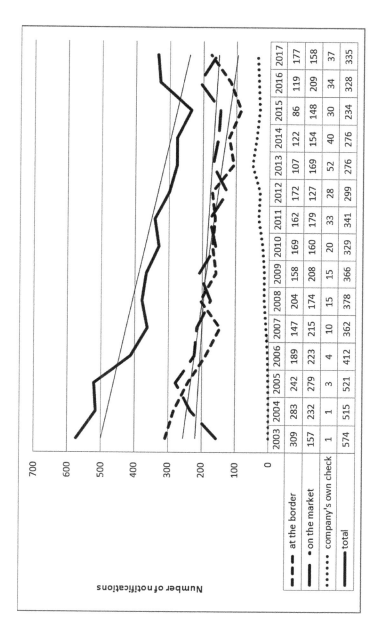

Figure 9.7 Source and number of notifications in Germany 2003–2017.

The current EU legislation allows for charging importers for monitoring. We must not under any circumstances give this away, and we need the EU to continue at the same rate as the UK (or *vice versa*) so that we can continue to afford to monitor; otherwise, I fear the impact could be catastrophic. We can already see how money for food monitoring is being reduced and our effectiveness as enforcers diminished.

REFERENCES

1. Statement from Environment Secretary Owen Paterson following the announcement from Findus and the Food Standards Agency about beef lasagne products, https://www.gov.uk/government/news/statement-from-environment-secretary-owen-paterson-following-the-announcement-from-findus-and-the-food-standards-agency-about-beef-lasagne-products [accessed August 2018].
2. Horsemeat scandal: Withdrawn products and test results, http://www.bbc.co.uk/news/world-21412590 [accessed August 2018].
3. Horsemeat: Shadow environment secretary on findings, https://www.bbc.co.uk/news/av/uk-21366544/horsemeat-shadow-environment-secretary-on-findings [accessed August 2018].
4. Why the delay? Findus left horse meat lasagne on shelves for SEVEN days, https://www.mirror.co.uk/news/uk-news/findus-left-horse-meat-lasagne-1595067 [accessed August 2018].
5. Horsemeat scandal: Can brands like Findus recover?, http://www.bbc.co.uk/news/business-21413966 [accessed August 2018].
6. Findus criticised for its handling of horsemeat crisis, https://www.theguardian.com/business/2013/feb/08/findus-crisis-horsemeat-communications-criticised [accessed August 2018].
7. Goodbye Findus crispy pancakes: Brand dogged by horsemeat scandal is to disappear, https://metro.co.uk/2016/01/31/goodbye-findus-crispy-pancakes-brand-dogged-by-horsemeat-scandal-is-to-disappear-5654449/ [accessed August 2018].

8. Tesco takes out full-page newspaper adverts to say sorry for horse meat burgers, https://metro.co.uk/2013/01/17/tesco-takes-out-full-page-newspaper-adverts-to-say-sorry-for-horse-meat-burgers-3355674/ [accessed August 2018].
9. Brands caught in the horse meat scandal should remember what happened to Perrier, https://www.moreaboutadvertising.com/2013/02/brands-caught-in-the-horse-meat-scandal-should-remember-what-happened-to-perrier/ [accessed August 2018].
10. Warning over rise in fake TSB text messages and phishing emails, https://www.horncastlenews.co.uk/read-this/warning-over-rise-in-fake-tsb-text-messages-and-phishing-emails/ [accessed August 2018].
11. TSB customers complain about fraud alert failures, http://www.bbc.co.uk/news/business-44156140 [accessed August 2018].
12. T. Nepusz, A. Petróczi and D. P. Naughton, *PLoS One*, 2009, **4**(8), e6680.
13. A. Petróczi, G. Taylor, T. Nepusz and D. P. Naughton, *Food Chem. Toxicol.*, 2010, **48**(7), 1957–1964.
14. A. Petroczi, T. Nepusz, G. Taylor and D. Naughton, *World Mycotoxin J.*, 2011, **4**(3), 329–338.

CHAPTER 10

Look What They've Done to My Song

In the words of the Melanie Safka song, *'Look what they've done to my song, ma, look what they've done to my song. Well it's the only thing that I could do half right and now it's turning out all wrong'*. Well I hope that's not the case. There have been major changes to food enforcement over the last 20 years with over 350 new regulations published since 2000, a loss of the good guys and the threat of a 'Brexit Armageddon'. Oh Ma, where have all the good guys gone and where will they be post-Brexit?

The good guys are those who uphold the law – food enforcement professionals who are now feeling under-valued. A tsunami of change has led to confusion. Austerity and Brexit have led to paralysis at the national level and local (food) authorities focussing on other responsibilities. Enforcement alone cannot eradicate food from fraud nor adulteration, its role to encourage compliance by offering a cogent threat of being caught for those who might consider the option to defraud or wilfully not comply.

10.1 SO MUCH CHANGE

Food law is, to say the least, complex and there is much lobbying by the industry and regulators for change. Currently, a review of the mechanism of delivery of food enforcement is underway, led by the regulators; unsurprisingly, the food industry is lobbying hard for more self-regulation and a change in focus. The food industry has conflicting priorities when it comes to food enforcement. They want to be left alone and be able to self-certify using third-party accreditation, something akin to the US model, but they do not want to have to share this information with the public at large, at least not in detail. Consequently, they are not keen on mandatory systems such as the FHRS, but they do want a level playing field with a cogent threat from enforcement for those who seek to gain an unfair competitive advantage by not following the rules. Historians suggest we should look to the past and learn from that before attempting to agree a future strategy. Previous chapters in this book detail the history. However, food adulteration is still noted and will continue to be noted. The introduction of early enforcers pounding the streets in the mid-1800s led to a dramatic improvement in food safety, and as a result, enforcement are defensive of any reviews which might lead to a reduction in resources, especially as the food sector continues to grow. If industry wants less intrusion, there is a real danger that enforcement will continue to shrink and the resources will not be available. They argue alongside advocates of both game theory and zero tolerance that this cogent threat of detection is needed today. The food industry argues that different strategies are available to support enforcement, that much has changed over the last 150 years and that enforcement must now be risk-based. Whoever wins the argument, the following key factors will impact on UK food enforcement's ability to be efficacious.

10.2 COMPLEX FOOD LEGISLATION AND A LACK OF HARMONY ACROSS THE GLOBE

During the early 19th century, when food adulteration was rife, two scientists led the charge towards better regulation of food safety, Frederick Accum and Arthur Hassall. The former wrote a best seller, *Death in the Pot*, which detailed how to identify non-compliant food, and the latter named and shamed non-compliant food producers and retailers. Such a strong response from the public to these two scientists and their work suggests to me that the public did not want to continue buying adulterated food. They wanted good wholesome food. Probably as a result, food enforcement became a high priority for regulators and politicians in the mid-19th century, and the 1860 Food Safety Act was enacted as a first attempt by UK Parliament to enforce food standards to ensure that the consumer received the food they thought they had purchased, *i.e.* unadulterated. This, and subsequent Acts, appointed local authorities as 'food authorities', and every food authority had to appoint enforcement professionals, such as scientists (public analysts) and Trading Standards Officers and Environmental Health Officers, to monitor the quality of food in their area.

Partly in response to the BSE crisis in beef, which was first noted during the 1980s, and the subsequent link to Creutzfeldt-Jakob disease (CJD) deaths in 1996, the EU attempted to co-ordinate and control food safety by introducing new legislation detailing how enforcement and regulators should be configured within member states. This standardisation of policies and systems led to FSAs being set up across the EU. These are non-government agencies and the intention is that they are free from political influence. The roles and expectations of the UK FSA are defined in the Food Standards Act 1999. This body has not taken away the power of the local food authority, which was defined within previous food safety Acts. Further detail on how the EU member states food control system must function was subsequently defined in EU198:2002 and EU882:2004.

As a result of the above legislation, food safety is 'enforced' at four levels:

- The WHO has overall responsibility for promoting international food safety and coordinating strategic direction at the international level. They promote, *inter alia,* the rapid sharing of information to secure food safety and public health around the world. They are assisted by the International Food Safety Authorities Network (INFOSAN), which brings together national authorities in European member states responsible for managing food safety emergencies.
- The EU Parliament has a responsibility for setting regulations/legislative standards and procedures in Europe.
- At the national level, FSAs are responsible for translating EU regulations and advice and guiding national governments

who are charged with adopting this into their own legislative frameworks. In the UK, this has subsequently become more complex as DEFRA and the Department of Health are involved as regulators and the FSA also enforce alongside the police for more major food fraud.
- At the local level, enforcement is undertaken by the local food authorities.

Therefore, regulators set the legislation and guidance to be followed by food companies. Food companies must abide by the law, and local enforcement enforces the legislation, thus monitoring compliance. However, not to confuse matters, but for the sake of clarity, it should be noted that the regulator, the FSA, does have an enforcement role on meat hygiene in abattoirs and for food crime.

10.2.1 Lack of Harmony in Legislation

The Beijing Declaration[1] was a failed attempt by the WHO to achieve cooperation, harmonisation and standardisation of legislative requirements at the international level. This aimed to ensure that producing companies and countries meet the exacting standards of all nations throughout the world, although not all nations have signed the agreement. This could have resulted in cheaper enforcement across the globe as importers would not have to test for compliance with their own standards; alas, inconsistencies in legislation remain. Finally, and perhaps most worryingly, the tests undertaken locally to demonstrate compliance with EU standards cannot always be replicated at the border to the EU, despite external accreditation of test houses. The largest food businesses are alert to these inconsistencies and have the intelligence as to which countries and suppliers tend to transgress. The smaller companies are more likely to receive goods 'on trust' in the absence of this knowledge and not have the wherewithal to afford testing, and unsurprisingly, the consignments sent for the smaller companies are more likely to fail to meet the stringent EU import standard.[6]

Despite attempts to harmonise standards, systems, approaches and legislation, the focus within the legislation is different. EU legislation defined in EU198:2002 and EU882:2004

Table 10.1 Costs of wine and sprit fraud to member states across the EU.

Member state	Lost through sales Euros (million)	Lost in excise duty Euros (million)
Spain	263	90
France	136	100
Italy	162	18
Germany	140	65
United Kingdom	87	197

both focus on risks to health, whereas the UK Food Safety Act 1990 demands that food is of the correct nature, substance and quality. The focus on quality does not necessarily include a risk to health. For example, diluted spirits may pose no threat to health but do defraud the consumer and likewise the government through reduced revenue collection. A report published in July 2018 by the EU Intellectual Property Office (EUIPO) states that, across the EU, spirits and wine fraud cost around €1.3 billion and details the revenue lost in some of the member states, as shown in Table 10.1.

Likewise, substitution of expensive foodstuffs with cheaper alternatives may defraud the consumer without threatening health, and products with special geographic protection status, such as Parma ham, may be substituted with other varieties without threat to consumer health. The recent substitution of beef with horsemeat prompted strong public and media reaction across the EU and in the UK and has shown that regulators need to be cognisant to public perception of the fraud and not just the health risk.

10.3 INCONSISTENT RESPONSE FROM ENFORCEMENT

Inconsistency in application of enforcement is rife across the EU. A simple analysis of RASFF notifications at the member state level shows that there are gatekeepers who detect more transgression than their neighbours because enforcement effort is not uniformly applied. Differing rates of sampling and expenditure on enforcement exist amongst EU member states and differences in strategies, such as charging for enforcement activity, particularly where produce does not meet the standard required, are also noted, despite this being encouraged through EU

Regulation. German legislation, for example, sets standards of sampling which are far higher than those of neighbouring member states and thus, perhaps unsurprisingly, the Germans are one of the key gatekeepers of EU food safety.

In the UK, enforcement is provided by local authority enforcement officers who are invited to work to national frameworks proposed by the central regulatory authority, the FSA. In the UK, there are around 400 local authorities, and asking them to work together and work at the same level of activity is fraught with difficulties. Each local authority is charged with focussing its resources in a manner which considers their local agenda (referred to as 'localism'), and this is defined by senior managers in agreement with their local politicians. This agenda will inform the local authority and drive its response to national strategies. Food may not even be on the agenda of the local authority if they prioritise other key issues such as care for an increasingly elderly population. Large food companies have lobbied government and complained at the lack of consistency between local authorities noted by large retailers and producers with premises in more than one local authority area. The large food businesses have asked for self-certification schemes based on management capability and previous performance to be taken into account when assessing the risk presented by large food manufacturers and retailers, arguing that the retailers themselves, through their own close monitoring of food manufacturers, already reduce the level of risk for the consumer. In his publication, *Reducing Administrative Burdens*,[2] Hampton was persuaded by these arguments and removed food from the list of priorities for enforcement, despite one million food-borne illnesses each year, 20 000 hospitalisations and 500 deaths.[3] Local authorities in the UK, responding to the call of Hampton and HM Treasury to reduce burdens on the food industry, cut their own resources and prioritised funds elsewhere, leading to less advice, sampling and inspection. Consequently, many local authorities now lack the resources to gather the necessary evidence to comply with Risk Based Regulation. *'Evidence demands visits and inspections, the very things Hampton wanted to reduce'*.[4]

In response to further calls from the food industry, local authorities have been encouraged away from the 'Home Authority Principle' of enforcement to a Primary Authority Partnership.

The Home Authority Principle of enforcement necessitated a lead for food enforcement being undertaken by the local authority where the products originated (or where the HQ of the large food business is situated). This was irrespective of local resources, and as a result, if local companies were not operating well and several complaints were made through other local authority trading standards and environmental health departments and referred to the Home Authority, the local authority may have insufficient resources as a result of the additional workload. No additional funding was available and the local authority would struggle to enforce. The later system, the Primary Authority Partnership, encouraged large and small food companies to select a local authority to provide environmental health and trading standards in return for a fee. Few large businesses have chosen this as they have in-house resources, and even fewer small food businesses have chosen this route as they do not have the resources available to pay. Web hosts may obtain their supply from smaller traders around the world, and as a result, may not get involved at all. Local authorities have continued to reduce the resources available and, in some cases, lack any resources to undertake any food enforcement.

10.4 RISK-BASED ENFORCEMENT

The EU structure separates risk into three categories: responsibility for risk assessment is based on scientific principles and lies with EFSA; risk management with the EU Directorate General for Health and Consumers (DGSanco); and risk communication for each member state lies with the FSA.[5] In an attempt to enable a more agile, targeted and effective response to enforcement, EU legislation calls for risk-based enforcement. EC/178/2002 defines risk as '*a function of the probability of an adverse health effect and the severity of that effect, consequential to a hazard*', once again only in regard to health and with a precautionary principle at heart. Horsemeat fraud would not feature as a key risk to health in this system.

EC882/2004 requires the identification of risk-based priorities and criteria for the risk categorisation of the activities concerned and the most effective control procedures – these are then included in Multi-Annual National Control Plans (MANCP) produced by the Competent Authority (the member state's FSA). Recital 13 of EC882/2004 states:

> 'The frequency of official controls should be regular and proportionate to the risk, taking into account the results of the checks carried out by feed and food business operators under HACCP based control programmes or quality assurance programmes, where such programmes are designed to meet requirements of feed and food law, animal health and animal welfare rules.'

No internationally agreed definitions of what this means in practice were provided at the time, and consequently, there have been many attempts to provide guidance and models for risk assessing food. Six different sets of guidance have been produced over the last decade in the UK:

- Local Authority Coordinators of Regulatory Services (LACORS) Trading Standards Risk Assessment Scheme[7]
- Reducing Administrative burdens: effective inspection and enforcement – Sir Phillip Hampton review[2]
- FSA food hygiene inspection rating scheme[8]
- The Roger's review of priorities for local authority regulatory services[9]
- Office of Fair Trading review of current trading standards risk assessment methods
- Risk Rating[3]

The above guidance all focuses on frequency of visits and omits the following:

- emerging intelligence;
- previous track record of the food company;
- capability of management of the food company;
- accreditation schemes, such as Red Tractor;
- perhaps most importantly from a health perspective, what level of risk is a satisfactory level at which no action is required; and,
- public perception of the issue, perhaps the most important issue for politicians.

This argument is used by the food industry as a key factor in the need to change the way enforcement works, and they are right to call for a change.

A great deal of work has been undertaken and time spent on risk-based enforcement, but is the holy grail the risk of transgression and the capability of enforcement to predict crime before it becomes a major issue, either to public safety or reputation. Police forces across the globe have worked hard on this issue, which requires intelligence and data. Enforcement would do well to work with the larger food businesses utilising their expertise and data.

10.5 WORLD-WIDE WEB OF FOOD SUPPLY

10.5.1 Increasingly Cosmopolitan Diets and a More Complex Food Supply Chain

Demand for food has changed. Consumers eat, on average, one in every six meals outside of the home and purchase more 'ready meals' for consumption within the home.[10] Three key factors have promoted a world-wide web of food supply: increasing demands from consumers for a year-round supply of seasonal products, a demand to satiate their more cosmopolitan palate and the desire for cheaper food. The latter, it has been argued, has led to more complex supply chains.[11]

Most non-compliance is noted in imported foods and most often detected at the border inspection post. In 2006, in the UK, imported food accounted for 21% of total food supply and 80% of the food alerts submitted to the EU *via* the RASFF database. Nine of the top 10 source countries were outside the EU.[12] Transitional economies, Brazil, Russia, India, China and South Africa (BRICS), are providing more of the world's food and they lead the way in terms of transgression too, as far as the UK is concerned.

10.5.2 More Complex Food Supply Chain

As the food supply chain (occasionally referred to as a web) grows, so too will economies in other parts of the world. This leads to transitional economies, which grow rapidly in response to demand. Transitional economies such as China are harder to police because legislation and resources for enforcement lag behind the market. Consequently, it can be difficult to enforce contracts and, as a result, they are not honoured.[13] It should be noted that China is doing a great deal to try to improve this situation but, inevitably, there is a lag between establishment of enforcement and market growth.

Complexity of supply requires more monitoring and policing, but of a different kind to that seen 150 years ago. Pounding the streets in the UK will have little effect on a business based in a transition economy across the other side of the globe. Larger food businesses realise this and use their resources for monitoring up and down the supply system. Research undertaken with German grain growers[14] demonstrated that producers would consider taking risks, and were more likely to do so – for example, use unpermitted chemicals or not follow allowed procedures – if they thought that tests could not identify the issue, or indeed that no monitoring was undertaken. The problem is more often encountered *via* direct suppliers to the UK market and not through the multi-national companies, provided they remain vigilant.

10.5.3 Is There an Insatiable Appetite for Cheaper Food?

It has been argued that over-complex supply chains in response to the desire to source cheaper food were one of the main causes for the adulteration of beef with horsemeat. According to the *Huffington Post* online newspaper, some retailers and producers were supplied with products containing horsemeat by Comigel, based in north-east France. Comigel instructed Tavola, its subsidiary in Luxembourg, to make the products. Tavola placed an order for the meat with Spanghero, in the south of France, who contacted a Cypriot trader, who subcontracted a Dutch trader. The Holland-based company placed an order with abattoirs in Romania, who sent the meat to Spanghero.[15] The more complex the supply chain, the more significant the risk of non-compliance.

10.6 AUSTERITY, INCREASING DEMAND AND REDUCING SPEND ON ENFORCEMENT

In 1874, there were 77 public analysts employed to fight food crime. By the mid-1950s, there were 40 laboratories employing around 100 public analysts. Since then, more than 22 laboratories have closed and the number of public analysts has fallen to less than 40. In the last five years, a number of public analysts' labs have closed in the UK, and five organisations remain in England and Wales.

An analysis of public finance using the Chartered Institute of Public Finance and Accountancy (CIPFA) statistics between the years of 2002–2008 has shown a fall of over 30% in the amount spent by local authorities on food enforcement.[16] The 2018 announcement from the National Audit Office is that local authority budgets have fallen by around 50%.

The FSA's own research shows both falling expenditure and an increasing need for enforcement. In England, the number of registered food premises, the number of food premises requiring interventions (actions) and the number of formal interventions (actions) all increased in 2010/11 compared to the previous year. However, the resources available fell.[3] Comprehensive spending reviews resulted in cuts across the UK amounting to 30% on local authority budgets. Consequently, three of the 10 enforcement laboratories in England announced closure during 2011 and many local authorities reduced the numbers of enforcement staff.

Spend by the UK FSA has also fallen by around 37% over the period 2008–2012;[17] see Table 10.2. In addition, the number of food samples taken by those enforcement staff has fallen by around 20% in the last three years, continuing a trend;[3] see Table 10.3.

One of the outputs of enforcement is RASFF notifications, and the 11th Annual report of RASFF[18] has revealed that, in 2012, there was a fall of 7.8% in the number of notifications compared with the previous year. This is the first fall in the numbers of

Table 10.2 Expenditure by the UK Food Standards Agency. Source: Consolidated Accounts.

Year	2007/8	2008/9	2009/10	2010/11	2011/12
Total departmental spending £ 000[a]	142 049	124 850	119 549	71 619	88 830
Reduced by %	0.00	12.11%	15.84%	49.58%	37.47%

[a]Total departmental spending is the sum of the resource budget and the capital budget less depreciation.

Table 10.3 Official samples taken by UK enforcement. Source: LAEMS annual returns.

Number of samples	86 324	105 556	92 122	78 653
Year	2008/9	2009/10	2010/11	2011/12

notifications, and even the key aspect to the EU's defence against adulteration, *i.e.* Border Rejection, saw numbers fall by 6%. Fortunately, there has recently been an encouraging announcement regarding an increase in funding available for the food crime unit; their budget is to be increased to enable them to detect food fraud.

10.7 THE ROLE OF THE MEDIA IN REPORTING NON-COMPLIANCE AND THE REACTION TO THESE REPORTS BY THE PUBLIC AND POLITICIANS

Attitudes towards risk vary across the EU, and therefore, a single approach is unlikely to work across all 27 member states. As previously discussed, risk of harm need not be to health alone, as fraudulent activity often targets wealth. For example, locally grown German asparagus was fraudulently exploited.[17] In year one, 90% of the products tested were not locally grown, the second year, around 50% and, in the third, around 5%. Local trader awareness of the new ability to enforce clearly had a major impact on an easy fraud, but how does this accord with EC178:2002, which lists risks to health? Is this a proportionate response from enforcement to a risk? This strategy does not accord with the various UK reviews of risk[2] and enforcement management, but it does accord with the German public's attitude towards reducing food-miles. The expense of monitoring this type of fraud and the number of people harmed suggest that this would be low on the radar of a risk-based system, if not ignored completely. Few, if any, enforcement labs in the UK have the capability to test locally grown produce. Industry looks towards enforcement to ensure that fraudulent behaviour does not impact on a 'level playing field' for all food companies, irrespective of size. Media discoveries of food adulteration attract the public's attention. Perhaps the media feed off the public's risk-averse nature and politicians are caught in the frenzy, especially if the public hold government responsible.

The media have an impact on public attitudes, particularly if the reporting is associated with a new issue or risk.[19] The recent substitution of horsemeat for beef has led to a strong public and political reaction despite reassurances from regulators at the FSA that this offers limited, or no, harm. Enforcement colleagues throughout the EU have been surprised by the reaction in the UK to this fraud, given the levels of risk from what some perceive as a minor fraud. This reinforced the belief amongst enforcement and regulators in the UK that the public, supported by media, have a zero-tolerance to non-compliance when it comes to food and expect people to be held accountable when transgression on a larger scale is identified, suggesting that the public are risk-averse where food adulteration is concerned. However, there is inconsistency, and the reaction to horse was much stronger than previous food scandals.

Professor Pat Troop noted in her review of the FSA's management of the horsemeat issue[20] that communication was key to the perception amongst stakeholders that the situation was being managed appropriately, and that this appeared to improve once a focus on communications with stakeholders, particularly through 'birdtable meetings', was established. This demonstrates that communication strategies need to be developed which:

- attempt to understand the perceptions of consumers;
- are more transparent regarding the risk analysis;
- communicate what is being done to protect (especially after an issue occurs), *i.e.* to make safe and prevent recurrence, including what work is being done with industry and enforcement; and
- adopt a differentiated targeted approach to communications, depending on who, what, where and how, with consistent messages.

Misinformation can be costly when food non-compliance is noted. The German *E. coli* in vegetables in the summer of 2011 demonstrates what can happen as unsubstantiated information is shared at an early stage. This 'set hares running' and led to accusations as to which country was to blame, having impacts on the economy of EU neighbours who struggled to sell their produce as a result.

Thus, the media has a vitally important role in communication and management of food issues, and food businesses, regulators and enforcers ignore the media at their peril.

10.8 RELATIONSHIPS BETWEEN THE KEY STAKEHOLDERS

Figure 10.1 shows the four key stakeholder groups in food safety; their relationships are developed in the following section.

10.8.1 Consumers

The consumer group (top circle) consists of the media and the public. The media both informs public opinon and interprets public opinon on a food issue, 'acting' on behalf of the public. Therefore, these two key stakeholders behave as a single entity.

10.8.2 Regulators and Enforcers

Throughout this book, I have used the terms regulators and enforcers (middle circle) almost as an interchangable entity; this

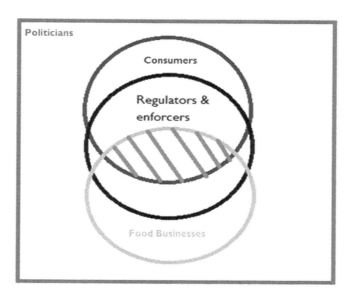

Figure 10.1 A schematic demonstrating the four key stakeholder groups in food safety and their relationships.

is because it is difficult to ascertain who has what role, and from the perspective of the consumer or food businesses, they are one and the same.

10.8.3 Food Businesses

The stakeholder group of food businesses (bottom circle) includes the food companies, representatives of the food trade, retailers, producers, suppliers and the countries within the food web. Each member of this group impacts on the others and their ability to provide food which meets the required legislative standard.

10.8.4 Politicians

Politicians are affected by the other groups and impact upon them through their decisions.

There is a dichotomy between ensuring risk management, which is the recommended approach to food safety management and targeting of resources, is based on both sound science and democracy. Democracy should ensure that the perceptions of the public are considered, but policy makers will also need to consider the scientific facts, including uncertainty. However, at a time of media frenzy, it is hard for policy makers – who include politicians who help agree those policies – to stand by their assessment, which was made in good faith and based on evidence available at the time. This challenge is yet to be convincingly met.[21] A good example to help explain this issue comes from a time that I worked with politicians over the fluoridation of drinking water. The scientific evidence suggests that, when fluoride is added to drinking water, it results in a reduction in tooth decay, especially for disadvantaged children who do not have good teeth cleaning regimes, and often no toothpaste nor brushes and a poor diet high in sugar. However, not everyone wants their drinking water to contain additional fluoride. And the politicians had to make a decision based on their perception of public opinion as well as the science. They chose to refuse to let the drinking water be fluoridated, preferring more targeted approaches for the disadvantaged children. It is worth noting that drinking water had to be fluoridated at source, and therefore, many properties would be provided with this water, not just the areas of deprivation. That was the dichotomy for politicians to consider in this particular case.

Local politicians also have a key role. They agree funding and priorities for local enforcement. Where ports exist in the area, enforcement may, and often does, struggle to divert funding towards the port as local politicians argue that this is not to the benefit of local people as imports leave the immediate area for onward transportation. Thus, local politicians argue that they are providing national protection.

The four stakeholder groups interact with each other. During the horsemeat crisis, there was a turning point at which harmony was achieved between the stakeholders when the FSA changed its incident response protocol from crisis management to communication with stakeholders through 'birdtable meetings'.

10.9 THE POTENTIAL FOR CONFUSION, ESPECIALLY REGARDING WHO IS RESPONSIBLE FOR FOOD SAFETY FAILURES

As stated earlier, there is the potential for role confusion between regulators and enforcers. Enforcement is provided by local food authorities and regulation from central government. With local authorities working to a 'local agenda' and national agencies such as FSAs responsible for food safety in the eyes of Ministers and the public, perhaps it isn't too surprising that there is confusion, especially when the 1860 Act which set up the food authorities made it a responsibility of local authorities to enforce, and this remains unchanged in current legislation. The FSA in the UK has the power to audit and inform local authorities where they fail to meet the required standard, and indeed, they have the availability of a 'Writ of Mandamus', which would enable them to take over a failing authority; but, despite damning audits, no action has ever been taken by the regulator. Perhaps a key issue is the fact that there are no clear guidelines on how to judge a failing food authority, and consequently, the challenge for those who uphold food safety, *i.e.* the national regulators and local authority enforcers, is how to demonstrate that they are efficient and effective when there are no standards by which to judge.

The FSA has responsibility for meat hygiene enforcement, licensing, inspecting and reporting on the meat processing plants around the UK and now also has the responsibility for the food crime unit.

10.9.1 Who Is Responsible for Food Safety?

Over the last decade, there has been increased emphasis on food self-regulation, based upon the ideas first promoted by Lord Robens[22] in his review of Health and Safety legislation. Enforced self-regulation relies on a mixture of state and company regulation. The State sets standards to which industry must comply and adopts a system of proportional monitoring, but primary responsibility lies with the company who must be able to demonstrate risk management systems that enable compliance or help the food business to use a defence of 'due diligence.' This is fundamental to the relationships between the regulator, food companies and enforcement and clarifies the roles and responsibilities of all parties. Self-regulation is gaining popularity amongst the larger food manufacturers and retailers and the regulator in the form of earned recognition. One of the key disadvantages of this system is that it is suited to larger well-informed and well-resourced businesses and is crucially reliant on businesses who aspire to self-regulate.[4] This moves the costs from enforcement to the food businesses. Some businesses may consider it more profitable not to comply with regulation and use enforcement checks as a form of quality assurance, particularly if enforcement does not transfer punitive costs from detecting the issue to the transgressor.[6]

Under Health and Safety legislation, it is the company that has ultimate responsibility and must comply with the law. When accidents occur, it is not the Health and Safety Executive that is seen as responsible for the failure. This is not the same with food safety. The public, media and politicians hold the FSA to account for failure. This situation adds to the confusion as to who is responsible and accountable for non-compliance.

10.10 POST-BREXIT ENFORCEMENT

It is hard to guess what the outcome will be in Brexit negotiations. Food standards will not be the highest priority for government negotiating key Brexit issues. The battle-bus used for the referendum (Vote Leave campaign) suggested that EU membership was costing the UK around £350 million each week, or around £18 billion each year. Considering that the money spent on food enforcement is less than 0.2% of that sum, it is unlikely to ever be high on the political agenda. Food standards aren't the highest priority for the public either; after all, they do not often react to food scares by calling for action, and they will not be overly concerned about the technicalities of food law negotiations – for example, whether the Americans can use a particular disinfectant which is banned in the EU to clean animal carcases. When it comes to negotiating trade deals between nations, food standards will not be the highest priority either. EU food laws are some of the most precautionary in the world, and that is likely to be challenged by others, if indeed the subject is even raised in negotiations at all.

10.11 ONE VOICE

The quality and safety of food is far too important to be just another priority in a long list of responsibilities that a government department or local authority must consider. It needs a voice at the national level, one that has government's ear, as was needed during the horsemeat crisis, an organisation that can hold government and the food industry to account. Local authorities typically spend less than 0.01% of their resources on food enforcement, while at government level, food amounts to approximately 0.1% of Department of Health budget, and this department has now been extended to cover social care too. How can food be a high priority for these organisations? Food needs, and deserves, a voice which can be heard, and therefore, a centralised body is a must. On this, the EU had it right – establish an arm's length national body, not a government department, and support it locally to provide local intelligence and knowledge. The central agency then can set regulations to advise government and industry, set protocols for checking for compliance, give freedom to industry to self-regulate through accreditation schemes, and oversee the auditors of industry and ensure transparency through mandatory systems such as the FHRS that provides information for consumers. Clarity of regulation systems and strategies for dealing with cross-border and web-based trading all need improving, and an organisation committed to achieving this is a must if we are serious about ensuring food safety. There are many models which can be successful, and what is needed is leadership to take enforcement forward.

Enforcement must move away from the old models and become more innovative to develop and use intelligence better to gain from a partnership with representatives of the food industry and work closely with countries that are shown to have high levels of transgression. A good example of this is EU-China-Safe, the work currently being undertaken by Chris Elliott *et al.* This is a system designed to standardise enforcement, learn from transgression and share innovation to stop further fraud and non-compliance. Data and intelligence use will form a backbone of future enforcement strategies, and the availability of information from tracking systems, 'Internet of Things' (the connection of Internet-based data systems which are fitted to every-day items such a fridges) *etc.*, will if shared, enable the efficient focussing of resources for both the food industry and enforcement.

Self-accreditation schemes should be encouraged and the regulator should have an overview of the quality of these schemes, and the accreditation organisations should be registered too. The outcomes of accreditation reports must be shared in a public manner, similar to OFSTED school reports.

Food businesses must be registered (as they should be now) so that enforcement knows where they are, and this registration needs to be replaced by 'a scheme granting permission to trade' so that businesses cannot trade without permission and a fee should be charged for the regulator to run such a scheme. The regulator will then be able to review the performance of the food businesses and the accreditation reports *etc.*

The UK should be proud of our record of food safety monitoring, which is of the highest standard and second in the EU due to port health, and this must continue post-Brexit to stop the UK becoming a dumping ground for sub-standard food.

REFERENCES

1. Beijing Declaration on Food Safety, http://www.wpro.who.int/foodsafety/documents/beijing_declaration.pdf [accessed September 2018].

2. Assessing our Regulatory System – The Hampton Review, http://webarchive.nationalarchives.gov.uk/+tf_/http://www.bis.gov.uk/policies/better-regulation/improving-regulatory-delivery/assessing-our-regulatory-system [accessed August 2018].
3. Statutory Monitoring Data – LAEMS, https://data.food.gov.uk/catalog/datasets/f4f6a8a5-d656-4eea-8746-1f121a94dff6 [accessed August 2018].
4. B. M. Hutter, *Br. J. Criminol.*, 1986, **26**(2), 114–128.
5. J. R. Houghton, G. Rowe, L. J. Frewer, E. Van Kleef, G. Chryssochoidis, O. Kehagia, S. Korzen-Bohr, J. Lassen, U. Pfenning and A. Strada, *Food Policy*, 2008, **33**(1), 13–26.
6. C. Yapp and R. Fairman, *Food Control*, 2006, **17**(1), 42–51.
7. Review of current components of risk assessment models that recognise central systems of businesses that operate across local authority boundaries, http://webarchive.nationalarchives.gov.uk/+/http:/www.berr.gov.uk/files/file37271.pdf [accessed August 2018].
8. Food Hygiene Rating Scheme, https://www.food.gov.uk/safety-hygiene/food-hygiene-rating-scheme [accessed August 2018].
9. Rogers Review – national enforcement priorities for local authority regulatory services, http://webarchive.nationalarchives.gov.uk/+tf_/http://bre.berr.gov.uk/regulation/reviewing_regulation/rogers_review/[accessed August 2018].
10. Food Statistics Pocketbook 2008, http://webarchive.nationalarchives.gov.uk/20130124042733/http://www.defra.gov.uk/statistics/files/defra-stats-foodfarm-food-pocketbook-2008.pdf [accessed September 2018].
11. Elliott Review into the Integrity and Assurance of Food Supply Networks – interim report December 2013, https://assets.publishing.service.gov.uk/government/uploads/system/uploads/attachment_data/file/264997/pb14089-elliot-review-interim-20131212.pdf [accessed August 2018].
12. Food: an analysis of the issues, The Strategy Unit January 2008, http://webarchive.nationalarchives.gov.uk/+/http:/www.cabinetoffice.gov.uk/media/cabinetoffice/strategy/assets/food/food_analysis.pdf [accessed August 2018].
13. W. D. Li, L. Gao, X. Y. Li and Y. Guo, *2008 12th International Conference on Computer Supported Cooperative Work in Design*, 2008, 841–845, [doi: 10.1109/CSCWD.2008.4537088].

14. N. Hirschauer and O. Musshoff, *Food Policy*, 2007, **32**(2), 246–265.
15. Horse Meat Scandal Has Highlighted Complex Supply Chain of Cheap UK Food, http://www.huffingtonpost.co.uk/2013/02/13/horsemeat-scandal-findus-supply-chain_n_2675465.html [accessed August 2018].
16. G. Taylor, *Forensic Enforcement: The Role of the Public Analyst*, Royal Society of Chemistry, Cambridge, 2010.
17. Food Standards Agency, Annual Report and Consolidated Accounts 2011/12, https://assets.publishing.service.gov.uk/government/uploads/system/uploads/attachment_data/file/247087/0036.pdf [accessed August 2018].
18. 2012 Report on Europe's Rapid Alert System for Food and Feed: Questions & Answers, http://europa.eu/rapid/press-release_MEMO-13-524_en.htm [accessed August 2018].
19. L. Frewer, *Toxicol. Lett.*, 2004, **149**(1–3), 391–397.
20. Review of FSA Response to Incident of Contamination of Beef Products with horse and pork meat andDNA, https://www.slideserve.com/onslow/professor-pat-troop [accessed August 2018].
21. P. van Zwanenberg and E. Millstone, *Political Q.*, 2003, **74**(1), 27–37.
22. Safety representatives, A Charter for Change, https://www.tuc.org.uk/research-analysis/reports/safety-representatives-charter-change [accessed August 2018].

CHAPTER 11

New Kids on the Block

11.1 THE NEED TO INNOVATE

Until now, there has been an EU-wide focus on risks to public health and ensuring the safety of the food chain from biological, chemical and physical risks. New intelligence gathered by Europol suggests that organised crime is involved in food fraud, which is estimated to cost the industry around $10 to 15 billion per annum (although the figure cannot be precisely estimated as the fraudsters seek to evade detection). They have suggested that allergenic ingredients are in the sight of these organised fraudsters, almond flour and powder being extended by the addition of ground peanut, for example. This is placing additional demands on our already overstretched food enforcement teams in both industry and governments. The detection of these crimes requires investment and strategic planning at an EU-wide level. The ultimate deterrent is the perceived ability of enforcement and food industry specialists to detect and punish non-compliance; therefore, they need to be able to demonstrate an ability to quickly identify and respond to issues as they arise and preferably be ahead of the game. To achieve this, there are calls for enforcement to become more agile, intelligence-led using more sophisticated analysis techniques and working in a similar manner to the police to better focus resources. The combination of increasing demand and falling resources necessitates innovative ways for working and achieving more with less. It is time to develop new strategies in the fight against food crime.

11.2 EXCITING NEW TECHNIQUES

The food industry is very keen on using more intelligent analytical techniques in the hope of finding a fingerprint for food to identify it and to prove its authenticity, in a similar manner to that used to identify people through their fingerprint or their DNA. This will be possible – if not now, then in the future – for raw ingredients. Indeed, we developed a system for the police drugs team to use to identify drugs of abuse (see section below), and this could be adapted now for ingredients, but I fear it is very unlikely to be the answer for composite and cooked foods.

Occasionally, new techniques emerge which get me excited. I am not sure if they awake a child-like curiosity or re-invigorate my scientific roots, but the techniques detailed below are fascinating:

- SNIF-NMR[R1]
  $^{13}C/^{12}C$, $^{2}H/^{1}H$, $^{18}O/^{16}O$, $^{15}N/^{14}N$ – Professor G. J. Martin at the University of Nantes has developed a process to provide an isotope analysis of foods to assess authenticity against the 'norm', *i.e.* establish a fingerprint for the food. This allows food companies to assess if a product or ingredient they buy has changed. If this information is shared, it could help enforcement check authenticity.
- Isotope ratio analysis δ^{18}Oxygen (for the origin of water used), δ^{13}Carbon and δ^{15}Nitrogen and δ^{34}Sulfur (for region of the soil), developed by Münster enforcement scientists (CVUA) in conjunction with Professor Schmidt at the University of Munich, can ascertain the geographical origin of food. This was used to prove that 'local' asparagus was anything but local and that the premium Skelt wine was in fact a fake with added, rather than brewed, fizz (carbon dioxide added from gas bottles). In both examples, prosecutions ensued.

  The above strategy is not available in publicly funded public analyst labs as it is expensive to develop and needs

many samples from enforcement agencies to justify the costs and development time. Our German counterparts at the state-owned CVUA estimate that development costs alone were over £250 000. Most local authority labs run at a deficit, and implementing expensive new technology is a distant dream. With a heavy heart, the need for this level of funding worries me about the future of public analyst labs and, indeed, food enforcement.
- *i*Knife
  a. the intelligent knife[2] is from the world of medicine and was developed by Dr Zoltan Takatsrial at Imperial College London. The knife cuts out tumours and healthy tissue during surgical procedures, and the smoke produced can be analysed in three seconds to identify if the tissue comes from a cancerous tumour. Research is ongoing to see if this technique can be modified and used for authenticity purposes to give an instantaneous result, perhaps at port of entry, for example, to identify if meat is as declared.
- DNA identification of meat species (based on Michael Walker's paper)[3]
  a. Initially (around the 1980s), methods for meat speciation were based on fatty acid profile analysis. These gave some success but were difficult to interpret in cooked food.
  b. Immuno-gel, or isoelectric focussing electrophoresis of soluble proteins followed, and this led to the development of the enzyme linked immunoassay (ELISA). ELISA methods are a popular analytical method of choice, owing to their ease and relatively low cost. They suffer lack of sensitivity below around 1% w/w and require careful handling at higher levels as the method is saturated by higher amounts of antigen. ELISA kits are generally sold as semi-quantitative. Electrophoresis techniques struggle to differentiate and therefore identify individual meat species in a mix, especially where one may be an adulterant at reasonably low levels, and even more so when the meats were cooked.

  Perhaps as a consequence, heat treated (cooked) meat products were more often adulterated than raw meat.

c. In the mid-2000s, the FSA funded the development of real-time DNA-based PCR assays for the specific detection of meat species. Techniques were established which detect minute traces of specific meat species such as duck, sheep, chicken, deer, horse and donkey. Real-time PCR generally offers greater sensitivity and specificity than PCR alone and is capable of being quantitative. DNA sequencing provides unequivocal identification of a species through reference to validated sequence databases. To validate the methodology, cooked and raw 'food samples' were spiked with horse or donkey muscle at sub-5% levels w/w and successfully tested for the presence of horse or donkey, demonstrating the applicability of the assays to food products. A set of standard operating procedures was provided for use by public analysts. The results must be interpreted by a scientist who is qualified and experienced to do so and one that is routinely involved in proficiency trials to establish their continuing professional capability. In addition, the lab must be accredited to ISO/IEC 17025. This system was used for the successful prosecution of horsemeat fraud in the recent scandal and shows great potential for development into other areas, not just meat.
- Portable spectroscopic (FTIR and RAMAN) and X-ray diffraction techniques to fingerprint raw ingredients
  a. Untargeted or fingerprint approaches to food integrity based upon spectrometry and chemometrics are established which identify non-compliance with the specified or usual composition. These have successfully been used for honey, olive oil and wine,[4] dairy products,[5] lipids[6] and rice flour.[7] In addition, easy to use portable infrared methods have been demonstrated for milk spiked with urea and synthetic milk. Mid-infrared (MIR) and near-infrared (NIR) data examined by soft independent modelling of class analogy (SIMCA), classification models and partial least-squares regression (PLSR) exhibited simplicity, sensitivity, low energy cost and portability.[8] Detection of animal feed spiked with melamine is also possible with these techniques.[9] Opportunities and challenges for untargeted analyses in official food control

have been discussed.[10] Critical workflow steps and 'good practice' for validation and reporting of non-targeted fingerprinting results have been published.[11] Similar methods have also been established within the forensic arena and produced excellent results for drugs of abuse detection in the hands of police officers. Consequently, this methodology is recognised by the UK Home Office Centre for Applied Science and Technology. Hampshire Scientific Services worked with police colleagues to develop portable FTIR (Fourier transform infrared spectroscopy) systems to enable rapid, on-the-spot testing for drugs of abuse by police officers. The system is now in use in music festivals, for example, saving police resources. These systems are ripe for development so that non-scientific personnel (*e.g.* auditors, quality control staff) in the food and feed industry can be trained to detect product adulteration and mislabelling of raw materials, and place an additional threat of detection to reduce the likelihood of fraud.

b. In his book, *Bad Ideas? An arresting history of our inventions*,[12] Professor Winston states that in '*every act of creation and innovation there is the potential for our undoing*'. He argues that technology used to improve can, and will, be abused, to our detriment. According to a report issued by the Cabinet Office in 2008, the food industry in the UK is worth an estimated £170 billion annually, so food crime can be very lucrative,[13] and the opportunity to use new technology may provide a quick profit on the necessary investment. The fraudsters have inside information and intelligence of their own which at least matches ours.

During the 1990s, a fraud came to light which featured a 'game of cat and mouse' between fraudsters and enforcement. Chicken sold *via* 'white vans' across the UK had similar properties; namely, difficult to cook, remaining pink irrespective of cooking times, often solid in texture and not pleasant to eat. Tests revealed high levels of organic phosphates and up to 50% added water. Much of this meat was unwittingly purchased by caterers, especially those who do not undertake their own monitoring of foods they buy. This is a lucrative market and one which has been identified by fraudsters on more than one occasion, with a total market value of around £200 million per annum. Whilst small volumes of water could be added to chicken to increase succulence, this practice must be declared to the consumer. The usual process of adding water is by tumbling the chicken breast in brine containing salts, sugars, possibly polyphosphates and milk proteins. This diffusion process is slow, and hence the amount of water that can be added to the chicken is limited. Change first occurred when fraudsters switched to multi-needle injection machines, and as a result, substantially larger amounts of water could be added in less than a minute. However, to retain the levels of water required by the fraudsters, around 30–50%, much stronger water retaining agents, hydrolysed collagen proteins, were needed. Research showed that these agents could retain 20–30 times the amount of water and that this would withstand a freeze–thaw process and cooking.

Now a lucrative crime was available with an excellent return. Brazilian and Thai chicken could be purchased, processed and frozen and then sold to unsuspecting caterers. Surveys in 2000 and 2001 showed this practice was occurring, prosecutions were taken and legislation changed as a result. Investment in new, more sensitive DNA profiling techniques was undertaken by enforcement, of the sort outlined above. This enabled detection of the collagen species, *i.e.* the species of meat, demonstrating that chicken was not the 'only meat' in the chicken breast portions.

The Dutch authorities identified five plants involved in adding extraneous water to chicken, prosecutions were taken for breaches of labelling rules, and in 2003, legislation was altered to require a declaration of the amounts of water/meat content and to ensure that no other species of meat was present in 'chicken fillets' *etc.*, *i.e.* that collagen from other species couldn't be used to increase the water content. Subsequent monitoring has shown that 'treated collagen' from other species (beef and pork) had been used to facilitate the addition of water to chicken fillets and that measures had been undertaken by the fraudsters to hide this collagen. This illegal practice was detected using the new knowledge and technologies being developed by enforcement at the time. The knowledge gained by enforcement was also used by the fraudsters, in a game of cat and mouse. Enforcement developed a new system to detect fraud, and the fraudsters found a way to hide the evidence; in this case, heating the collagen (from other species) in an attempt to destroy the structure and then adding some DNA from chickens.

The melamine in baby milk fraud was another example of science being used to defraud. This fraud demonstrated an inside knowledge of the current practices for testing milk and it 'fooled' enforcement, which was only using one test to assess the apparent milk content. Most enforcement scientists would view both the adulteration and the practice of relying on one test with horror, but clearly this was happening in China, if not elsewhere. If we are successful in identifying fingerprint techniques, will we become reliant upon them in a similar manner to the Chinese authorities? Will they be as unique as a person's fingerprint?

11.3 HOW DO YOU USE INTELLIGENCE TO DEFEND AGAINST FOOD FRAUD?

The definition of a holy grail might be something desirable that is unobtainable. Some argue that getting the large food businesses to share intelligence across the globe is a holy grail. Many in the food enforcement arena have confided in me that they do not think it is possible. It is happening in other areas, particularly the police, so perhaps a look at what they are achieving will help.

The challenge for all in enforcement has been the prediction of crime and the use of innovative techniques that stop crime occurring rather than detect it afterwards. Sir Richard Mayne, Joint First Commissioner of the Police of the Metropolis, said (in 1829):

> *'The primary object of an efficient police (force) is the prevention of crime: the next that of detection and punishment of offenders if crime is committed.'*

This (prevention through prediction of crime) is one of the five key principles behind the National Intelligence Model (NIM),[14] developed in the UK and used by police across the world. The key principles are as follows:

  I. Target scarce resources in areas most needed.
  II. There are three levels of crime (which may intertwine):
   - Local crime
   - Cross police area crime
   - Serious organised crime.
  III. Don't respond – predict.
  IV. Engage the public.
  V. Place the data in the public domain to enable them to use the information.

Nearly all of the above could be agreed by the food industry, regulators and enforcement, the only stumbling block being the last bullet on placing data in the public domain.

Researchers suggest that, as data emerge, there is a point where it is possible to predict an emerging threat and to take action to stop it.

There is a moment when a driver becomes aware that another vehicle isn't going to stop and give way when it should. Police drivers are trained to look at the wheels of cars at junctions to help them predict earlier if a vehicle starts to move, enabling them to use this information to predict a road traffic collision before it happens.

From a police-crime perspective, there are more crimes committed in areas where alcohol and people mix. Crime maps show the statistics which support this[15] and enable the information to be provided to the general public so that action can be taken by police and public. The next step for police to consider is how to target resources effectively to reduce that likelihood. This may be an early or timely police presence in areas where higher levels of alcohol-related crimes are noted, or strategies to reduce alcohol consumption by promoting moderate drinking. In Weymouth, police and publicans are using breathalysers to tackle the culture of 'pre-drinking' in which a growing number of people consume excessive amounts of shop-bought alcohol at home before arriving in the town centre.

The broken windows and zero tolerance theories to crime, as adopted by the ex-Mayor of New York (Giuliani), are based on similar principles to the NIM, namely engaging the public in reporting crime and swift action by the authorities to repair any damage in order to demonstrate that crime is not accepted and will not be tolerated. One young New York graffiti artist was found in tears after he saw his 'work' immediately washed off a subway train before it was even dry. The simple message, graffiti will not be tolerated on New York subways.

Police colleagues have often talked about the crime continuum, which if left unchecked, can start with teenagers spraying graffiti and lead to damage, then minor drugs offences and culminate in a life of crime. It is argued that a zero tolerance approach in an area can stop this sort of crime escalation in its tracks or break the continuum. I suspect that many of us have judged a new area we visit from the amount of graffiti and damage that can be seen. This does translate into food businesses where an accepting culture in an organisation may lead to a higher prevalence of 'misdemeanour' and possibly fraud. If the food business works hard to ensure that everyone associated with it is aware that there is a zero tolerance to anything but the best, then there is a reduced likelihood of fraud.

11.4 ANALYTICS EARLY DETECTION OF EMERGING DATA

The key issue is where does the intelligence come from to predict future crime? A while ago, I worked with police intelligence officers and data analysts. We talked about 'Dashboards' visualisation of emerging data. In the case of a car, the dashboard consists of dials which tell the driver what is happening and interpret the emerging data. One of my previous cars had an intelligent system fitted and, on one occasion, produced an audible warning that I should pull over when safe to do so. I didn't need to understand what the oil pressure should be or what it was, I needed analysis by a 'smart system' (often referred to as analytics) to tell me what action to take to prevent further damage. Hampshire enforcement agencies (police, trading standards and citizens' advice bureau) used a dashboard system to predict crime and take advance action, in some cases warning residents. For obvious reasons, I cannot provide a copy of the live system here, but shown in Figure 11.1 is a food mock-up based

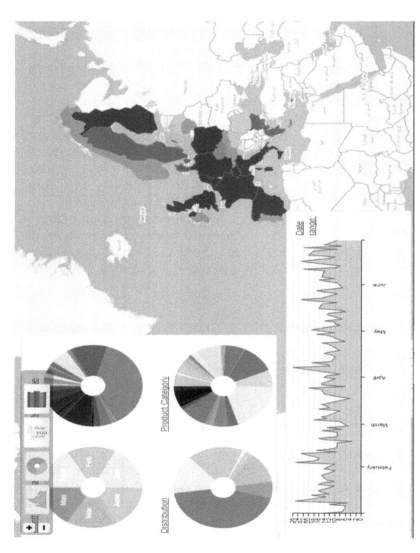

Figure 11.1 A live visualisation of emerging data based on RASFF data for the purpose of predicting crime and trends.

on RASFF data. On the live system, clicking on the image enables further analysis of the data to identify trends.

A plot of previously identified issues against time (Figure 11.2) shows a point at which there was a rapid increase in the frequency of detection across the EU, an emerging trend. By having the right dashboard and analytics (the discovery, interpretation and communication of meaningful patterns in data), it will be possible to predict an emerging trend at an early stage.

An intelligence continuum is outlined in Figure 11.3.

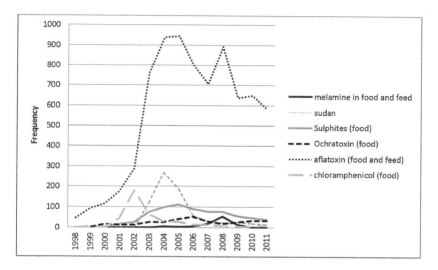

Figure 11.2 The frequency of EU detections of a variety of contaminants 1998–2011.

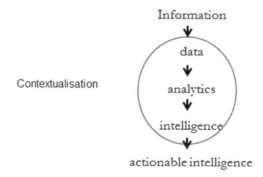

Figure 11.3 Schematic demonstrating an intelligence continuum.

Several pieces of information become data when they are combined, and an analytical approach will help us understand the data. The key factor is the contextualisation based on professional knowledge as this converts simple data into intelligence and leads to actionable intelligence, something that can be acted upon. This approach is based upon that used by trauma medics in war zones such as Afghanistan. Several medics work alongside each other receiving a trauma patient and sharing developing information to rapidly assess the best way forward for the patient, deciding which clinical strategy is best and which steps to take first *etc.* This has been successfully developed and is in use in emergency rooms in hospitals around the world and has saved the lives of many people involved in trauma, stabbings, shootings and road traffic incidents, for example.

Enforcement officers use '5×5×5 intelligence recording systems' to provide information, for example, that a particular vehicle/van has been seen in an area which belongs to 'known people'. This, combined with complaints in the same area of poor quality fish being sold door-to-door and the knowledge that fish is often targeted by fraudsters,[16] could lead to actionable intelligence that a known fraudster has moved across an enforcement boundary border to sell fish and that subsequent monitoring may show this to be fraud. This may, upon further analysis, relate to an annual event such as a market or harvesting cycle. As a result, action can be taken to prevent crime in future by sharing the information with the public in the area so that they are not fooled into buying 'bargain products', combined with a timely enforcement presence in following years. This is similar to the police example outlined above and the example of local asparagus sold in Münster.

So far, the examples I have given focus on fraud at a local level, but efforts are being made to widen this. Police colleagues share data through organisations such as Interpol. Some of the research being undertaken will facilitate this wider approach.

11.5   WWW DATA SHARING RESEARCH

The University of Singapore and Tsinghua University of China have combined to provide some exciting developments in live analysis of breaking news which can be interrogated

(NExT++).[17] Of the many news sources that they monitor, when a fresh item is released, the relevant site picture flashes. When watching in real time, lights flash constantly and, by using key word searches, the news coming in can be filtered and results targeted. The data would allow food companies to interrogate the system for key events or issues emerging from Asia.

The University of Southampton is part of the SOCIAM research team. SOCIAM is the theory and practice of social machines, which has, at its heart, an open source of web-based information in order to engage people to solve problems and mobilise resources (people, technology and information). This massive collaboration across the globe has produced a couple of exciting possibilities for food fraud; namely, use of emerging signals to predict a trend and provenance.

11.5.1 Use of Emerging Signals to Predict a Trend

The brain processes thousands of signals each day, so an algorithm developed by SOCIAM researchers at the University of Southampton has helped neuroscientists in Southampton University Hospital to assess when signals from the brain are linked. This information can then be used as a potential way to assess the recovery of the brain, following trauma for example. This research has been further developed by the SOCIAM research team to form a Macroscope (Figure 11.4) – a monitoring system which enables the team to look for emerging trends across the world-wide web for a particular issue. They facilitate searching of data for trends of interest and the ability to identify when a trend starts.

This will facilitate monitoring of emerging trends in discussion, for example tweets, and the monitoring of breaking news using systems such as NExT++. In this way, the public can share concerns regarding food safety. As a food scare emerges, the system would be able to identify it at a very early stage.

11.5.2 Provenance

We are all familiar with the notion of provenance for art, which is used to demonstrate that a Renoir painting is genuine. The first step is to enquire about its provenance – its back-story.

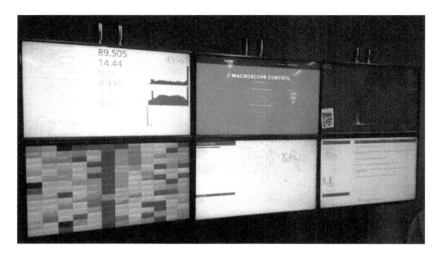

Figure 11.4 An example of a Macroscope – a monitoring system which enables a team to look for emerging trends across the worldwide web.

In the case of food, products from a specific designated origin carry a premium value, as do some brands, so it is just as essential to know the provenance. This back-story or provenance became even more important in the recent horsemeat scandal when, in order to successfully prosecute the case, provenance of the food needed to be proved to the high criminal standard.

The Oxford English Dictionary defines provenance as:

'the fact of coming from some particular source or quarter; origin, derivation the history or pedigree of a work of art, manuscript, rare book, etc.; concretely, a record of the passage of an item through its various owners'.

The World Wide Web Consortium defines provenance as follows:

'Provenance is a record that describes the people, institutions, entities, and activities, involved in producing, influencing, or delivering a piece of data or a thing in the world'.

Therefore, it is possible to produce a data system which will allow the provenance of a food to be accurately 'mapped' and monitored, as well as identify any weak links within the process. As the provenance maps of some foods are complex, systems such as analytics are used to monitor the weak points and look for changes in the system. Professor Luc Moreau and his team at Southampton University are mapping school meal supply chains for Hampshire County Council and assessing what analytic tools need to be developed to allow real-time analysis. This can easily be developed to map foods across the globe in a similar manner to that already undertaken for timber to demonstrate, amongst other things, the sustainability.

11.6 CHALLENGES TO USING INTELLIGENCE

On a world-wide scale, police and enforcement agencies already share data and intelligence. This may change following Brexit but my fervent hope is that it does not. One of the key difficulties that has been noted in all aspects of data sharing and intelligence is 'sensitivity'. As you would imagine, crime statistics can be very sensitive and personal. In the crime map system, referred to earlier, data are anonymised so it's not possible for the public to ascertain that a particular crime was committed at 123 Acacia Avenue, for example. Nevertheless, the anonymised data are publicly available and allow some targeted action in a particular location to be taken.

The challenge for the food industry is to share data and intelligence in an open manner as they have some of the best contextual knowledge and data sets, which will enable their data to provide actionable intelligence. The holy grail is to develop an intelligence system which maximises the effectiveness of resources. The key question remains will the food industry rise to that challenge?

The days of a sufficient food enforcement 'presence' to stop crime in the first place are no longer viable given the size of the supply chain, the lack of money governments across Europe are prepared to allocate and emerging economies with a lack of harmony in legislation. 21st century food enforcement needs sophisticated intelligence systems that rely on the involvement of industry and the public, as well as an industry that recognises that a cogent threat needs to be tackled by proactive, successful enforcement. Therefore, the sharing of data and intelligence nationally, internationally and globally is of paramount importance. This is one holy grail we need to find.

Large organisations have the skills and knowledge to help contextualise information emerging from horizon scanning, including for example, opportunities to defraud, weaknesses in supply chains and areas least likely to be monitored. If food businesses are to help regulators and enforcers, they will want something in return. Currently, they call for 'Earned Recognition', which recognises the current performance of a food trader's operators, acknowledges their management system and targets the scarce enforcement resources away from these companies on the proviso that food companies will deal with the issues and notify enforcement/regulators when they find them. In many cases, these are already in existence. Enforcement already has knowledge as to who are the strong and well-managed food companies in their local area and use this information to target their resources elsewhere. Unpublished research undertaken by the author on behalf of FLEP has shown that around 25% of enforcement samples show non-compliance. This level of non-compliance is not reflective of the whole of the food industry but demonstrates targeting by enforcement across the EU. However, current risk-based systems call for high risk business premises to be visited once per annum by enforcement officers, and the performance in this regard is monitored through annual returns from enforcement to the regulator (FSA). Both industry and enforcement think it is time for a rethink on this policy – time to be more intelligence focussed.

11.7 SUPERHEROES OF FOOD ENFORCEMENT

11.7.1 The Procrustean Bed of Food Enforcement

In Greek mythology, Procrustes offered a bed to travellers. If the guest didn't fit, he amputated limbs or stretched them during their sleep. This fable gives rise to the concept that one size fits all. The original Food (Safety) Act 1860 took a one-size-fits-all approach when identifying who would analyse food samples to ascertain compliance with the law, namely public analysts. It has worked very well for around 125 years, and many might think they were the superheroes of food enforcement. However, complexity and a lack of investment have led to many asking if this is a viable option anymore.

Like the original, the Food (Safety) Act 1990,[18] gives public analysts a key role in the analysis of food for public consumption. Public analysts have been the superheroes of food enforcement in particular; they are the only forensic scientists with a legal right to present evidence in court without establishing their credibility and track record. All other forensic scientists will be cross-examined on their suitability as an expert witness; not the public analyst. They are charged with guiding the courts where legislation on food does not exist. This may be, for example, interpreting the results and whether or not this is a satisfactory result or likely to be to the prejudice of the consumer. The qualification that gives them this right is the Mastership in Chemical Analysis (MChemA), and the UK is the only member state which operates this system. The Germans have a similar qualification which enables food scientists to demonstrate they are competent, and members of the food industry are also qualified to this level, not just enforcement scientists (as in the UK). Many local authority public analysts are woefully underfunded and lack access to the latest technology, critical if they are to continue the fight against food crime. They try valiantly to use the skills and equipment they have, but a lack of investment, coupled with a resistance to collegiate working, hampers development; this frustrated the FSA, which wants to see the latest techniques used. Consequently, the FSA has further developed its list of official control laboratories[19] to include labs with expertise in raw milk, shellfish analysis and marine biotoxins. As we find more complex techniques, the list will continue to grow and include a wider range of scientists who are not qualified as public analysts. Is it a problem if we move away from a Procrustean bed approach, one that has served us so well?

EU legislation rightly demands that competent authorities must be accredited to EN ISO/IEC 17025 to demonstrate that the quality of their work fits the bill. More recently, the Forensic Science Regulator has pursued the same approach, insisting that forensic science labs must be accredited for all the analysis they undertake. The problem is interpretation of the rules as some member states expect very specific accreditation schedules; for example, melamine in milk or melamine in biscuit. I have called for caution in this area in the past. During the melamine crisis, no labs in the UK were accredited for testing baby milk for melamine. Fortunately, one or two control labs were able to undertake the analysis and used their flexible scope to enable them to demonstrate that the method was under control and thus, where necessary, prosecutions could follow. In the UK, this would be permitted on a temporary basis until the next round of assessments (assuming the lab continues to undertake this work). In Germany, they have a different approach to accreditation as they accredit techniques not specific foods. For example, food would be checked for melamine by using the HPLC process, and that is all the schedule would detail. In Hungary, colleagues tell me that it is necessary for the specific analyte in the specific food to be accredited *before* the work can be undertaken and that, for new tests, this takes around 6–12 months to obtain. As a result, a quick response to the melamine crisis was not possible in Hungary. Complex accreditation schedules add to the overhead costs of the labs, and many would-be food enforcement labs might seek to avoid this cost. They also aid the fraudster who knows that prosecution may be more difficult where accreditation rules are so stringent.

11.7.2 Is it Time for a Rethink?

There are both public and privately funded public analyst labs in the UK, and Professor Chris Elliott has thrown down the gauntlet to government funded public analysts to get their act together so that they are on a more viable financial foundation and incorporate new techniques. They have failed to achieve this so far, and consequently, new techniques are even further away. Earlier in 2018, they signed an agreement to start to work together, but investment and reorganisation are still a long way away. Some of the privately funded public analyst labs do have better access to new technologies. Eurofins, a multinational lab, offers public analyst services across the UK and the EU, including I understand, being the sole provider of these services in one member state. I, for one, respect Eurofins. Therefore, one option open to the UK is to solely rely on one or more privately funded public analysts. I have been told by friends in government circles that this is not the preferred option, and in his review of food safety,

Chris Elliott stated that a government funded lab was necessary. I spoke with a previous chair of the FSA who agreed with me that one private sector and one public sector public analyst lab could, and should, cater for Britain. There are only a handful of public analysts working in the UK, and the number of trainee public analysts has fallen over successive years, leading to more retiring than entering the profession. The current breakdown is as follows: England – 6 local authority laboratories (10 public analysts) – 3 private laboratories (PASS) (7 public analysts); Scotland – 4 local authority laboratories (7 public analysts); Wales – 1 private company (2 sites) (3 public analysts); a total of 27 public analysts currently working in the UK.

The FSA is looking for options around scientific enforcement. The meat hygiene service, which they administer, gives them an option to enforce meat, and in conjunction with the police, they are now tackling food fraud through the food crime team. They could lead enforcement science, bringing the public analysts from local authority ownership under their umbrella. They have shied away from this in the past because they do not want to take the risk and associated costs. They would prefer a local authority to take the lead.

The quality of the work is not in question and, in my experience, it has always been exemplary, but if we are not going to halt the decline in the existing model, then we must look for a different strategy. It is sad and ironic that police forensic science is going down the same road, as total spending in this area continues to reduce year on year. As a result, research and development has suffered, resulting in the National Police Chiefs Council, formerly the Association of Chief Police Officers, calling on forces to centralise a pot of money (from the savings obtained by not funding the now closed forensic science service) to be set aside for future research.

The Procrustean bed approach served food enforcement well until the 1980s, when science and food (supply and additives) both increased in complexity. Competitive tendering resulted in business practices which led to division and competition between labs and a downturn in investment and training. Our German counterparts cannot believe that UK analysts compete and do not share expertise. This intensity of competition has only increased since the inception of the FSA, and over the last decade and a half, they have not been able to find a solution. In fact, it feels further away now. This approach has damaged food enforcement science and change is desperately needed. Sample numbers continue to fall to catastrophic levels, and yet the UK has an excellent pedigree and is a gatekeeper of EU food enforcement.

11.8 HOW DO WE MOVE FORWARD?

We have four options:

- Change the law to remove the need for public analysts to be involved in food enforcement. I fundamentally disagree with this proposal; given the pedigree and quality of work already in place, this would be akin to throwing the baby out with the bath water. In my opinion, it would simply result in less competent people providing cheaper analysis, mirroring some of the sometimes accredited labs seen in distant shores where results provided for certification for export purposes to the EU cannot be verified by competent accredited enforcement labs across the EU.
- Let the private sector provide public analyst services – this model already exists and works. Even our referee analyst is in the hands of the private sector. Industry involvement with privately owned labs has encouraged them to research and develop new techniques which are lacking in the public sector. It is hard to argue against although some, including me, have expressed a preference for an independent government-owned lab which doesn't work for the food industry, to provide a greater degree of independence and avoid a monopoly of provision.
- Leave it all to chance and hope for the best. Often the preferred option in government circles, especially during times of austerity, although the consequences are inevitable: the numbers of labs will diminish, investment will continue to fall and labs will close, leaving the private sector to pick up the mantle.
- Grasp the nettle and establish a national service. Elsewhere, I have suggested that national coordination of food enforcement is needed with local authorities handing over the

resources they currently employ to a central body (as has been seen with the meat hygiene service). This must include a government-run public analyst.

Sadly, the current system continues to fail; we shouldn't strive for the easy option and just leave it all to chance. In my view, the private sector model works, but tendering will continue, leading to non-existent food enforcement at the local authority level. If the FSA and DEFRA really want an independent service which can develop techniques to be used by enforcement, one that they can develop to their needs, then they must grasp the nettle and, to use the vernacular, put their money where their mouth is, stop prevaricating and develop a centralised food enforcement service which includes a UK public analyst lab. So the dilemma for government is whether or not to continue with the original Procrustean bed approach or have a centralised service, or let the private sector run food enforcement. Eventually, local authorities will do less and central government will need to take over.

REFERENCES

1. The SNIF-NMR® Concept, http://www.eurofins.com/food-and-feed-testing/food-testing-services/authenticity/snif-nmr-concept/ [accessed August 2018].
2. Cancer surgery: Tumour 'sniffing' surgical knife designed, http://www.bbc.co.uk/news/health-23348661 [accessed August 2018].
3. M. J. Walker, M. Burns and D. T. Burns, *J. Assoc. Public Anal.*, 2013, **41**, 67–106.
4. L. E. Rodriguez-Saona and M. E. Allendorf, *Annu. Rev. Food Sci. Technol.*, 2011, **2**, 467–483.
5. L. Xu, S.-M. Yan, C.-B. Cai, Z.-J. Wang and X.-P. Yu, *J. Analy. Methods Chem.*, 2013, 201872.
6. A. Koidis and M. T. O. Argüello, *Lipid Technol.*, 2013, **25**(11), 247–250.
7. L. Xu, S.-M. Yan, C.-B. Cai and X.-P. Yu, *Food Anal. Methods*, 2013, **6**(6), 1568–1575.
8. P. M. Santos, E. R. Pereira-Filho and L. E. Rodriguez-Saona, *J. Agric. Food Chem.*, 2013, **61**(6), 1205–1211.

9. S. A. Haughey, P. Galvin-King, A. Malechaux and C. T. Elliott, *Anal. Methods*, 2015, 7, 181–186.
10. S. Esslinger, J. Riedl and C. Fauhl-Hassek, *Food Res. Int.*, 2014, **60**, 189–204.
11. J. Riedl, S. Esslinger and C. Fauhl-Hassek, *Anal. Chim. Acta*, 2015, **885**, 17–32.
12. R. Winston, *Bad Ideas?: An arresting history of our inventions*, Bantam Press, London, UK, 2010.
13. Food: an analysis of the issues, The Strategy Unit, January 2008 (Updated & re-issued August 2008), http://webarchive.nationalarchives.gov.uk/+/http:/www.cabinetoffice.gov.uk/media/cabinetoffice/strategy/assets/food/food_analysis.pdf [accessed August 2018].
14. The National Intelligence Model, http://www.intelligence-analysis.net/National%20Intelligence%20Model.pdf [accessed August 2018].
15. Crime Map reported in June 2018 for the area Strand and Whitehall, https://www.police.uk/metropolitan/00BK17N/crime/ [accessed August 2018].
16. Oceana: 1 in 5 seafood samples mislabelled, http://www.foodnavigator.com/Market-Trends/Oceana-finds-economically-motivated-adulteration-of-seafood [accessed August 2018].
17. NEXT++ website, http://www.nextcenter.org/ [accessed August 2018].
18. Food Safety Act 1990, http://www.legislation.gov.uk/ukpga/1990/16/contents [accessed August 2018].
19. Multi-Annual National Control Plan for the United Kingdom, Appendix M, https://www.food.gov.uk/sites/default/files/media/document/ukmulti-nationalcontrolplan2013-2018.pdf [accessed September 2018].

CHAPTER 12

It Ain't What You Do It's the Way That You Do It

It's abundantly clear that a system originally designed to ensure food safety 150 years ago now needs revitalising. Food is no longer produced and adulterated locally. A world-wide web of distribution, coupled with scientific opportunities to alter what we eat, the way it is produced and the source, provides plenty of opportunity to defraud, necessitating improved ways of working and not just more resources for the professionals involved. Enforcement, like every public service, calls for more resources to increase effectiveness. However, in a time of austerity, can it justify the additional spend? It isn't what you do so much as the manner in which enforcement and regulators act.

A system of charging and competitive tendering, and successive efficiency initiatives introduced throughout the latter part of the last century, reduced enforcement's capacity to monitor food composition, a fate avoided by those undertaking microbiological testing. This has led to reduced investment in training and technology and in scientific services responsible for providing up-to-date expertise in support of enforcement. Dr Brian Iddon MP stated at a House of Commons review '*that the number of enforcement officers has diminished at what some would describe as an alarming rate since the 1950s. In 1959, 150 public*

The Horse Who Came to Dinner: The First Criminal Case of Food Fraud
By Glenn Taylor
© Glenn Taylor 2019
Published by the Royal Society of Chemistry, www.rsc.org

analysts worked out of 45 laboratories, in 1997, there were 32 laboratories'.[1]

In 2000, with the discovery of BSE, a new focus on food safety arrived. Across the EU, food enforcement was coordinated, remodelled and defined. Organisations were set up and given responsibility for aspects of food enforcement; for example, risk assessment was managed and overseen by EFSA, risk management by DGSanco and regulations by each member state's FSA.

Comprehensive spending reviews since 2008 have led to further reductions in expenditure on food enforcement. A 37% fall in spend by the UK FSA[2] and a 32% fall in samples submitted to public analysts for testing led to three laboratories closing in 2011, leaving only a handful of laboratories in the UK. In 2018, this number has shrunk still further, as three more labs have announced closure. There are no prescriptive levels for sampling in the UK, and as a result, local authorities striving for lower costs consider reducing the numbers of samples to as low a level as possible, and in some cases, none. In 2013, the UK sampled at a rate of two samples per 1000 population. In Germany, there is a prescribed rate of sampling of five samples per 1000 population. In the UK, this has reduced yet demand for enforcement is increasing. In England, the number of registered food premises and the number requiring interventions and actions all increased in 2010/11. However, the resources available fell,[2] as did expenditure on enforcement.

Consumers eat, on average, one in every six meals outside the home. They demand a more cosmopolitan diet and purchase more 'ready meals' for consumption within the home.[3] This necessitates a world-wide web of food distribution which facilitates non-compliance and fraud. Around 78% of our food in the UK is sourced from within the EU, with the remainder sourced from regions beyond EU regulatory enforcement.

There is no single benchmark to judge the performance of enforcement in regard to food safety. The media and public look for non-conformance by the food industry and use this to judge government effectiveness. Enforcement uses a system to communicate when a food or feed fails to comply with EU legislation – the RASFF.[4] In 2006, 80% of food alerts on the RASFF database related to the 23% of foods sourced outside the EU, showing that the majority of issues now relate to food sourced from outside

EU enforcement. The RASFF system is not designed to facilitate comparison of enforcement, but an analysis of the RASFF notifications (RASFF Portal) has led to peer-reviewed papers that compare using the number of detections noted by a member state. Italy, Germany, UK and Spain lead the way.[5] Since 2004, despite falling sampling numbers, the UK has improved its position in the league table. These data also show a reduction in the numbers of detections of transgression 'on the street' and an increase of detections, particularly in Germany and the UK, in the numbers noted by large food businesses, suggesting an opportunity to work more closely with industry in particular as resources are reduced. Consequently, it can be argued that, despite falling resources and increasing demand within the UK, the enforcement team have successfully maintained their position when compared to other member states across the EU.

The EU is only as strong as its weakest link. Freedom of movement of food within the EU offers a threat to food security if one member state does not maintain an adequate focus on border enforcement. An analysis of ports across the EU[6] revealed a 129-fold difference in the effectiveness of enforcement at the ports, with the Netherlands and Belgium being gateway ports into the EU. The Annual Report of RASFF[4] has revealed that, in 2012, there was a fall of 7.8% in the number of notifications. This is the first fall in these numbers and also in the EU's defence against adulteration. Border Rejection saw numbers fall by 6%. In 2012, for the first time, the UK produced the most RASFF notifications in the EU (15% of the total): 517 original notifications, matching those made by Italy. Thus, the UK maintains a strong position.

12.1 HOW COULD ENFORCEMENT IMPROVE?

Several opportunities exist for improving food enforcement, which include:

- Decide the mission: does food enforcement (FSA) want to be seen as responsible for food safety or should it adopt a similar stance to that of the Health and Safety Executive and others. Then publicise a 'business consumer and regulator pact'.

- Improve strategic leadership: the local agenda for food enforcement needs to change. Better centralised strategic coordination is required which sets clear expectations and responsibilities for local authority enforcement, perhaps through national boards like the National Trading Standards Board. Local knowledge is vital to an enforcement service but, given the complexity of food enforcement, the local agenda may be best served through a more regional framework and the technological challenges through co-ordinated centres of excellence.
- Closer working with industry: share information with industry, particularly the largest organisations that have expertise, resources and technical capabilities. Develop 'earned recognition' systems which are thorough and can be used to reduce the need for enforcement.
- Involvement of the public – the wisdom of the crowd: encourage whistleblowing and provide information for the public to help enforce standards, *e.g.* make the FHRS mandatory. Engage with the public to educate, communicate and learn.
- Improve learning: learn from other organisations outside the food industry, and look at audit and counter fraud measures. Separate media and crisis management in the event of a crisis. Learn from major failures: undertake risk and reliability analysis, avoiding tick-box ratings which simply restate the previous thoughts – 'inflated confidence of man'.
- Think like a criminal: consider opportunities to defraud using waste products, *e.g.* horsemeat, leather and melamine.
- Encourage and share innovation: technological and business, *e.g.* shared services across government.
- A focus on resilience: future failure is inevitable if we continue to expect zero risk/failure. Food enforcement will need to identify issues and respond quickly.

REFERENCES

1. Public Analysts Service, June 2009, 11am, https://publications.parliament.uk/pa/cm200809/cmhansrd/cm090602/halltext/90602h0004.htm#09060250000002 [accessed September 2018].

2. Food Standards Agency, Annual Report and Consolidated Accounts 2011/12, https://assets.publishing.service.gov.uk/government/uploads/system/uploads/attachment_data/file/247087/0036.pdf [accessed September 2018].
3. Cabinet Office 2008: Food: an analysis of the issues, The Strategy Unit, January 2008, (Updated & re-issued August 2008), http://webarchive.nationalarchives.gov.uk/+/http:/www.cabinetoffice.gov.uk/media/cabinetoffice/strategy/assets/food/food_analysis.pdf [accessed September 2018].
4. RASFF, The Rapid Alert System for Food and Feed 2012 Annual Report, https://ec.europa.eu/food/sites/food/files/safety/docs/rasff_annual_report_2012_en.pdf [accessed September 2018].
5. A. Petróczi, G. Taylor, T. Nepusz and D. P. Naughton, *Food Chem. Toxicol.*, 2010, **48**(7), 1957–1964.
6. G. Taylor, A. Petróczi, T. Nepusz and D. P. Naughton, *Food Chem. Toxicol.*, 2013, **56**, 411–418.

CHAPTER 13

Who Have We Invited to Dinner Next?

13.1 SO WHO IS COMING TO DEFRAUD YOU NEXT?

Look around your dining table; to continue the metaphor, it is highly likely that the fraudster is already there. Yes, it is possible that you may be defrauded by an outside company selling copies of your products, as seen in the French wine frauds of Chapter 4 or even the Head and Shoulders fraud, as reported in the *Express* newspaper on 4 December 2014. Apparently, more than 160 bottles were seized by Trading Standards Officers following complaints by customers in the west of England. Each bottle was found to be a hazardous fake and a very good fake too, as the manufacturers themselves confirmed it was hard to tell the difference between the original bottles and the fake, although the contents were very different and allegedly resulted in injury to some customers.[1] This sort of fraud is less frequently encountered than an 'inside job', someone in your supply chain taking advantage of a weakness and their own intelligence and opportunity. The chances are the fraudster is sitting at your table. Of course, you may not know them directly. The longer your supply chain, the more likely they are there, lurking, waiting for the opportunity and not on your radar.

The Horse Who Came to Dinner: The First Criminal Case of Food Fraud
By Glenn Taylor
© Glenn Taylor 2019
Published by the Royal Society of Chemistry, www.rsc.org

13.2 IS THE PURSUIT OF CHEAPER FOOD TO BLAME?

Sir Malcolm Walker, the Founder, Chairman and Chief Executive of Iceland, the frozen food company, suggested that the horsemeat fraud was simply a product of local authorities driving down the amount they are prepared to pay suppliers for the ingredients for school meals. According to him, they got what they deserved. Undoubtedly, in order to provide school meals at the price required by central government, there has been a sustained drive towards lower costs from local authorities. However, not all local authorities simply focussed on cost without streamlining processes and monitoring value for money. The authority I worked for regularly tested school meals at all stages of manufacture, shortened supply chains using local companies and regularly visited the suppliers, forming close partnerships and sharing the results of our monitoring. However, I am aware that we were perhaps only one of two local authorities undertaking this monitoring. I have some sympathy with his views however, as some of the food industry has focussed on driving costs down to a minimum in order to compete on price. I once heard a technical manager from a food business saying that defence against food fraud was likely to add to their costs and did they have to have such monitoring systems just to ensure they were less likely to be the target of fraud. The simple answer is, if you are prepared to accept fraud and only pay lip service to monitoring, then fraudsters will come looking for you. I understand that margins are tight, especially for some large food businesses that focus on providing cheaper food, but you must send out a loud and clear message that you check and will not tolerate fraud.

I listened to a presentation by one of the senior food technology managers from Marks and Spencer (a company I admire) about how they eradicated *Campylobacter* from their chickens, a problem that has vexed retailers for some time. The person in the audience next to me, from another very large food business, confided that the margins of his employer are so tight that if, by adding the extra checks recommended in the presentation, they would add 5p to the price of a chicken, they would simply lose sales to a cheaper competitor. Their market is that competitive. Yes, perhaps Marks and Spencer customers will pay a little more and, therefore, the company can afford more vigorous checking systems. So Sir Malcolm, the CEO of Iceland, I do have some sympathy with your views; if food businesses are not prepared to pay for monitoring, then they will get what they deserve – fraud – as will those local authorities who do the same. Maybe food businesses need to hold their nerve and prices and not buy from the cheapest supplier, as this may lead to reduced monitoring and checks, as the suppliers feel they have to cut overheads. Some customers will, of course, seek cheaper food, but they have a right to be protected by the law, which holds food companies responsible for the produce they sell.

13.3 SHOULD WE THE CUSTOMER BUY DIRECTLY FROM THE OTHER SIDE OF THE WORLD?

We have seen that the longer and more complex the supply chain, the more likelihood of a fraudster getting involved, particularly when there seems to be less threat of detection. As a result, some may ask, if UK food businesses cannot rid themselves of fraud, if we should cut out the middle man and buy directly from a supplier based on the other side of the world. Whilst this may shorten the supply chain, *caveat emptor* (buyer beware). You do not know the supplier if the only contact you have is *via* the Internet, and you may not be adequately protected by consumer laws in your country, depending on the contract and how easy it is to trace the supplier. Personally, I stay with the local outlet; there is more chance of them being held to account if things go wrong, and this should be incentive enough for them to maintain sufficient monitoring.

13.4 FOLLOW THE MONEY

Professor Lisa Jack from Portsmouth, suggests using forensic accounting techniques to monitor food fraud; she has tried to evaluate the amount of food fraud occurring. She suggests it is impossible to give an accurate figure as fraud, by its very nature, is not in the open and not easy to assess. Work carried out by PKF Littlejohn and Portsmouth University suggests that food fraud amounted to 5.5% of expenditure, *i.e.* over £11 billion, in 2014 for the Footsie-listed companies alone. She suggests that systems of forensic accounting which follow the money, *i.e.* the profits made by each supplier in the chain, will show where the largest profits are being made, and if there are significant improvements in any one supplier, look here for the possibility of fraudulent behaviour. From a company perspective, look at your weakest products, *i.e.* those where you have the least information on your suppliers, especially where the supply chain is longest.

13.5 SO WHAT PRODUCTS ARE NEXT?

I hear you ask what products are next. Some would say how do we predict the answer to the impossible question, but frankly, the answers are in front of us:

- High value products. The products of designated origin which attract a premium price; for example, Parmigiano cheese or Aberdeen Angus steaks, oils such as olive oil, wines and spirits.
- Same old same old. Look at the RASFF notifications. They aren't changing significantly and, in fact, have not changed that much in 150 years, other than the quantity of adulteration. It must be recognised that food adulteration is nowhere near as rife as it was back then, but the same foods are adulterated; why, because they are the ones we buy the most.
- A twist on a current theme. Horsemeat wasn't a new fraud, just a modification on a theme, as we have seen substitution of lamb and beef many times in the past.
- The soft target. Food supplements in the widest sense of the description. These are a soft target, especially when supplied from across the globe or through 'friends'.
- A shortage of a product. Especially when it's the new food fad, or perhaps because a chef on TV has suggested we all use this ingredient or for other reasons it is in short supply. A friend who is the technical manager of a large food retailer told me the story that a few years ago a TV personality recommended the use of peppers in salads and she gave several recipes. His company was inundated as demand went literally through the roof, and they simply could not get

enough peppers. All was good until they carried out a check and found pesticides. They went to their supplier in China and asked why. In years of working together, they had never had a non-conformance; everything was as it should be and the relationship with the supplier was excellent. The supplier's response was, '*I didn't want to let you down, orders went up exponentially and I sourced peppers on your behalf, from local suppliers*'. Of course, the supplier didn't carry out any research or checks, and the results could have been very difficult for the retailer if they were not on the ball.

- Waste chemicals, which can disguise test results, or areas of limited testing, such as melamine.
- Issues in other areas. Melamine was noted in pet food 18 months before it was found in baby milk.

Now, if you were expecting me to name products, you will be disappointed (and unrealistic – that was never going to happen). Look in the areas above, and you will find fraud if you look hard enough.

13.6 IS THERE ANY GUIDANCE TO STOP THIS?

There is a great deal of guidance around on how to protect your industry from fraud. The Global Food Safety Initiative (GFSI) has recently published *Tackling Food Fraud through Food Safety Management Systems*.[2] This builds on previous work by GFSI and the key message is just 'get started' on food fraud systems within your company. The British Retail Consortium has excellent guidance too, as does the FDA with their food defence plan builder system.[3] This system enables the company to build a plan and is well worth a look if you are starting the process or want to check your rationale. Whatever system you follow, you should encompass other systems, and not build this as a stand-alone process, linking, for example, waste monitoring and disposal, HACCP systems, *etc*. Once you have risk assessed areas, prioritise the higher risk areas first. There is a great deal of free help available; you could start with online advice from John Spink Michigan State University, for example. My advice is that you already have systems to protect your products from a food safety viewpoint; expand these by looking at the opportunities a criminal might look for. Where are the weaknesses in your processes? Good luck.

13.7 BACK TO THE BEGINNING OF MY CAREER

My first sample as an apprentice chemist working in a public analyst lab back in 1975 was to analyse an alleged aphrodisiac called Spanish fly [Cantharidin (Scheme 13.1)]:

Cantharidin is secreted by the male blister beetle (sometimes called Spanish fly) and given as a copulatory gift to the female so that the female can use it to protect her eggs after fertilisation, as it wards off predators. It is a burn agent and toxic in larger dosages, and somehow it was identified as an aphrodisiac. I was disappointed, as a young 17-year-old analyst, to find I only had to confirm the chemistry, not the effectiveness of the claim. By the way, nosiness got the better of me and a quick check shows Spanish fly is available online should anyone wish to buy it. The sample I checked passed the test. Back then, we looked at samples like that. It was known as street surveillance and led to intelligence, but now you would be hard pressed to find a similar sample in any of the public analyst labs as enforcement, following lobbying by the major food businesses, are asked to gather intelligence first, before sampling. Of course, none of us had intelligence to suggest that horse was being used as an adulterant for beef or, in some cases, virtually a replacement. As a result, if it wasn't for the street surveillance by the Irish, horse would not have been found, at least not for a considerable time, and the whole shift in emphasis may never have happened.

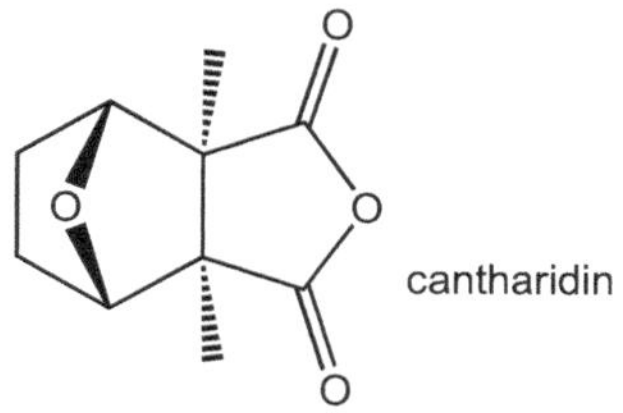

Scheme 13.1

13.8 WHAT WILL HAPPEN POST BREXIT?

That has to be the $64 000 question. In short, who knows? One week there is a press release suggesting cheaper food as the tariffs are removed. The next, a House of Lords European Union Energy and Environment Sub-Committee committee report, 'Brexit: food prices and availability',[4] which outlines the differences of opinions from no tariffs leading to a price rise of 3.8% on food, due partly to the increased costs of monitoring at the border, to current tariffs with the EU which are, on average, 22%.

95% of the world's trade is through the WTO. This system states that preferential treatment should not be shown to one nation or group of nations. In other words, tariff free should be for all, not one or two nations. However, a nation can negotiate free trade with others. This does mean that, if the UK were to waiver tariffs on EU food, it would need to do the same for all WTO nations. The government has stated that it wants free trade with the EU. If there is no agreement then world trade rules would apply.

In a nutshell, it looks to me like food prices will increase post-Brexit. Will the government want to fund additional checks at the border as this may restrict movement of food? I doubt it.

The next major issue will be the freedom of movement of food across borders. There is a great deal of lobbying from the food industry representatives for frictionless borders. Currently, delays at border inspection posts can amount to five days and this, they argue, increases the cost of food. If we agree to continue with free movement, we will need to align our border checks with the rest of the EU or risk traders finding the easiest import route to the EU, which could be Rotterdam or Southampton depending on how the UK decides to operate. Frictionless borders will necessitate a different mechanism for testing at the border which will result in more cost to either the government or food industry. The UK is one of the gatekeepers of food safety – the border inspection posts ensure we remain a gatekeeper. Brexit negotiations will have repercussions, and what the outcome will be, at this stage, can only be speculated.

Now the issue for me isn't where the next fraud will come from, but will it be detected. Will we maintain resources, and given a world-wide web of supply of products, will our systems cope? And, then, what impact will Brexit negotiations have? I just hope that enforcement and regulators have enough lobbying power to ensure a vibrant team remains and that it has sufficient resources to offer support to consumers who chose the web as their supplier.

REFERENCES

1. Warning over fake bottles of Head & Shoulders containing dangerous 'sex change' chemicals, https://www.express.co.uk/news/543627/head-and-shoulders-shampoo-fake-britain-warning-chemicals-dangerous/amp [accessed August 2018].
2. Tackling Food Fraud Through Food Safety Management Systems, https://www.mygfsi.com/news-resources/news/news-blog/1396-tackling-food-fraud-through-food-safety-management-systems.html [accessed August 2018].
3. Food Defense Plan Builder, https://www.fda.gov/Food/FoodDefense/ToolsEducationalMaterials/ucm349888.htm [accessed August 2018].
4. Brexit: food prices and availability, https://publications.parliament.uk/pa/ld201719/ldselect/ldeucom/129/129.pdf [accessed August 2018].

Subject Index

abattoirs and slaughter-
  houses 4–5, 8, 9–10, 12,
  165, 171
accreditation 8, 69, 142, 180,
  199, 200, 202, *see also*
  training
    self-accreditation 180
Accum, Fredrick Carl 46, 48,
  51, 122, 123, 124, 163
Actus Reus 22, 89, 121
additives *see* adulteration;
  supplements
ADHD 94
adulteration (with additives)
  45–59, 93–4, 121, 213
    definition 52, 54,
      62, 64, 93
    historical perspectives
      45–59, 62, 121–3, 163–4
    olive oil 71–2
    public attitude 50–1
    verdict of 57–8
Adulteration of Food and
  Drink Act (1860) 48, 121
aflatoxin 153
Ainsworth, Rod 87
alarm (burglar) principle
  132–3, 135

alum (Leucogee) 45, 47, 122
Amazon 92–4
amphetamine 112
analysts and analysis *see*
  emerging data; laboratory
  services
Anti-fraud platform, European
  (AAC-FF) 62–3
antibiotics 90, 134
asparagus, German 66–7, 173,
  184, 194
athletes and sportpersons
  105–11
attention deficit hyperactivity
  disorder 94
audit 37–9, 134
austerity 43, 93, 97, 116, 142,
  155, 158, 162, 171–3,
  202, 205
authenticity and integrity
  (food), Elliot Review *see*
  Elliot Review

baby milk, melamine 21, 22,
  68–71, 117, 189, 199–200
banking
    crisis of 2008 139–44
    TSB bank fraud 150–1

Barnes, John (of FSA) 35
BBC 148, 150
  'Victorian Bakers'
    documentary 50, 124
Beaumeunier, Virginie 75
beef (cow meat)
  Bovine Spongiform
    Encephalopathy (BSE)
    and 21, 164, 206
  Polish sold as
    British 14, 28
  substitutes for 1–2, 4, 6,
    9–17, 147–8, 149
    chicken in 60–2
    horse see horsemeat
    prosecutions/court
      proceedings 9–17
Beijing Declaration 113, 114,
  116, 165
Belgium 17
  Dutch eggs and fipronil
    62–3, 64
benzene in bottled water 149
Benzo Fury 112
Bioterrorism Act (2002) 91–2
Blackwell, Thomas 47–8
bleaching of fish 78–80
Boddy and Moss (case) 9–12,
  87, 100
borders
  EU 90, 104, 113, 114, 115,
    116, 207
    freedom of move-
      ments across
      207, 216
  UK 158–9, 170, 216
bottled water, benzene in 149
bought in good faith 55, 96
Bovine Spongiform Encephal-
  opathy (BSE) 21, 164, 206

Brazil, see also BRICS
  economies
  chicken 188
  fish 180
bread, historical perspectives
  44, 45, 47, 49–50, 51, 57,
  122, 124
breaking news 77–8, 194–5
Brexit 104, 116, 158, 162,
  178–9, 180, 197, 216–17
BRICS economies 70, 104, 170
British Retail Consortium
  (BRC) 37, 38, 134, 214,
  see also United Kingdom
broken windows theory to
  crime 190–1
brokers 4
  audit 37–8
  registration as
    businesses 134
Brooks, Stephanie 24
Brown, Gordon (Chancellor)
  140, 141
BSE (Bovine Spongiform Encephalopathy) 21, 164, 206
budgets see funding
burglar alarm principle 132–3,
  135
businesses/companies/industry
  134, 134–6
  banking crisis (2008)
    and 139–43
  brokers registered as
    businesses 134
  closer working with 208
  consumers and the roles
    and responsibilities
    of 24–8, 134–6
  government and 134,
    141, 142–3, 179, 216

businesses/companies/
industry (*continued*)
    horsemeat scandal-
        involved 26, 147–9, 150
    intelligence 191, 197, 198
    as key stakeholders in
        food safety 176
    registration 91, 92, 180
    regulators and regu-
        lations 88, 146, 176
    zero tolerance 28–9
buts, Turkish 95

Cameron, David 86, 148
Campylobacter, chicken
    90, 211
cantharidin 215
capsicum *see* peppers
Caribbean red snapper 80
cattle *see* beef; milk
cayenne pepper 48, 123
centres of excellence 33, 208
cheaper food, demand
    171, 211
cheese, red 48, 123
Chemische und Veterinär-
    untersuchungsamt (CVUA)
    67, 184, 185
chicken, *see also* eggs
    1990s–2008 scandal 21
    beef pies containing 60–2
    Campylobacter 90, 211
    chlorinated water for
        washing 90
    water addition 187–9
chicory, coffee with 47, 58
children
    melamine in baby
        milk 21, 22, 68–71,
        117, 189, 199–200

school meals 97, 124,
    135, 197, 211
'Southampton Six' and
    ADHD in 94
chilli powder, India 82, 94
China 170–1
    data sharing research
        194–5
    EU-China-Safe 180
    melamine in baby milk
        21, 22, 68–71, 117, 189,
        199–200
    peppers containing
        pesticides 214
chlorinated water for washing
    chicken 90
chocolate 48, 70, 95, 123
citric acid 79
CJD (Creutzfeldt–Jakob
    disease) 164
cocoa 48, 123
coffee 46, 48, 123
    with chicory 47, 58
collagen 188–9
collective intelligence *see*
    sharing of data/
    information
colours and colouring (and
    dyes) 94
    chilli powder 82, 94
    fish 79–80
    'Southampton Six' 94
Comigel 171
commodity prices *see*
    prices
communication 132–3,
    see also media
    with the public 27
    systems 109, 133
companies *see* businesses

Subject Index

competition (laboratory services *etc.*) 35, 202
    tendering 34, 201, 205
compliance 162
    lack (non-compliance) 56–8, 114, 127, 163, 170, 174, 198
    tests/checks/ensuring/monitoring 104, 113, 117, 140, 165, 178, 179, 198
confectionery 48, 95, 111, 123
consumers and customers 24–8, 175, *see also* public
    buying from other side of the World 212
    demand *see* demand
    Elliot Review 24–8
    protection world-wide 89
    synthetic food and 52
control *see* defence
costs of fraud, *see also* funding; prices
    EU, wine and sprits 166
    global 124
    UK 125
court proceedings (prosecutions and sentences)
    Chinese melamine scandal 79
    clarity for those involved 98–9
    Crown Court *see* Crown Court
    custodial sentences *see* custodial sentences
    due diligence defence 54–5, 89, 94, 95–7, 187
    EU 16–18
    five elements demonstrating fraud 3, 54
    horsemeat *see* horsemeat
    Magistrates Court 2, 22, 55, 87, 95, 100, 143
    mitigation plea 54–6
    new guidelines 87–9
cow *see* beef; milk
Creutzfeldt–Jakob disease 164
criminal, thinking like a 120, 208
crisis management 41–2, 177, 208
Crispy Pancakes (Findus) 148
crop failures 130
Crown Court 100, 136
    Southwark 9–16
Cruickshank, Tom 140
custard powders 48, 123
custodial (prison) sentences 13, 18, 100, 101
    China 70
customers *see* consumers
CVUA (Chemische und Veterinäruntersuchungsamt) 67, 184, 185
cyanuric acid 68

dashboard system (early detection of emerging data) 191–2
data *see* emerging data; sharing
Davies, Owen (judge) 15
deaths
    DNP 108, 109
    E.coli O157-contaminated meat 96–7
    Kratom (*Mitragynine speciosa*) 110–11
    melamine 69

defence (control and prevention/protection) 132–6, *see also* enforcement
    guidance 214–15
    strategies 132–6
    weakest link 116, 136, 207
DEFRA (Department for Environment, Food and Rural Affairs) 22, 39, 40, 141, 147, 165, 203
demand 130, 170, 206
    increasing 130, 170, 171–3
        cheaper food 171, 211
    shortages 213–14
Department for Environment, Food and Rural Affairs (DEFRA) 22, 39, 40, 141, 147, 165, 203
Department of Agriculture, Food and the Marine (Ireland) 6, 16
Department of Health (Ireland) 6
Department of Health (UK) 39, 113, 141, 147, 165, 179
deregulation of banks 140, 141
detections
    Brexit and 216–17
    capability/likelihood of 132
    changes in methods of 123
    early detection of emerging data 191–4
    laboratory services *see* laboratory services
    list of food types in order of frequency of 122
    RASFF and member state comparisons of number of 207
DGCDRF (Direction Générale de la Concurrence, de la Consommation et de la Répression des Fraudes) 75
diet pill, DNP 108–10, 116–17
2,4-dinitrophenol (DNP) 108–10, 116–17
Dino and Sons 14, 15
Direction Générale de la Concurrence, de la Consommation et de la Répression des Fraudes (DGCDRF) 75
distribution, vulnerability during 129, 206
DNA testing 185–6
    equine (horse) 6–7, 9, 16, 17
    pork 6, 7
DNP (2,4-dinitrophenol) 108–10, 116–17
documentation and record-keeping 5, 14–15, 88
    inadequate or falsified 5, 10, 11, 12, 13, 175
Doll, Sir Richard, and toxic oil syndrome 73
Domaine Ponsoet and Laurent Ponsot 76–7
drugs and medicines 111–12
    definition of medicine 105
    illicit, mimics 111–12
due diligence 54–5, 89, 94, 95–7, 187

Subject Index

Dunnes Stores 7
Dutch, the *see* Netherlands
dyes *see* colours and colouring

early detection of emerging data 191–4
EC 178/2002 64, 88, 98, 168
education *see* training and education
eggs, fipronil in 62–6
electrophoresis techniques 185
ELISA (enzyme-linked immunoassay) 185
Ellard, Ray (of FSAI) 16
Elliot Review (and Professor Chris Elliott) 22–43, 86, 125, 133, 200, 201
  eight pillars
    audit 37–9
    consumer 24–8
    crisis management 41–2
    government support 39
    intelligence gathering 29–30
    laboratory services 31–7
    leadership 40–1, 40–1
    zero tolerance 28–9
  success assessment 42–3
emerging data/signals
  early detection of 191–4
  in predicting trends 195
EN ISO/IEC 17025 186, 199
EN ISO/IEC 17925 69

enforcement (incl. food safety) 136, 162, 162–217
  banking crisis and lessons to be learned 141–3
  Elliott Review enabling *see* Elliot Review
  four levels of 164–5
  future *see* future
  inconsistent application 166–7
  innovation and new strategies 180, 183–204, 208
  key stakeholder role of enforcers 175–6
  local *see* local authorities
  performance 23
    assessing 206
    improving 207–8
  reduced expenditure/investment 171–2
  risk-based 141, 168–70, 198
  specialist teams 121, 133, 183
  superheroes 198–202
England
  Food Safety and Hygiene (England) Regulations (2013) 87, 88
  public analysts 201
  reduced spending on enforcement 172
  Sale of Food Act (1582) 45
enzyme-linked immunoassay (ELISA) 185
equine DNA testing 6–7, 9, 16, 17

Escherichia coli
   meat contamination with type O157 96–7
   vegetable contamination 174
EU-China-Safe 180, *see also* European Union
Eurofins 6, 35, 37, 67, 200
European Anti-fraud platform (AAC-FF) 62–3
European Food Safety Agency (EFSA) 63, 66, 90, 94, 168, 208
European Union (EU) 1, 16–18, 69, 72, 113–16
   borders *see* borders
   Brexit and 104, 116, 158, 162, 178–9, 180, 197, 216–17
   EN ISO/IEC 17025 186, 199
   EN ISO/IEC 17925 69
   fish bleaching and colouring 78–80
   Food Standards Agency (EFSA) 4, 90, 94, 164–5, 168, 169, 206
   frequency of detection across 193
   gatekeepers of food safety 151, 154–8
   genetic modification 90
   Intellectual Property Office 166
   laboratory services 34, *see also* Eurofins
   RASFF *see* Rapid Alert System for Food and Feed system
   regulations and law
      EC 172:2002 52
      EC 178:2002 64, 88, 98, 168
      EC 198:2002 165
      EC 882:2002 98
      EC 882:2004 164–5, 168, 169
      enforcement 164, 168–70, 173, 174
      harmony and consistency problems 89–90
      new guidelines and regulations 88
   sports supplements 105–6, 106, 109
Europol 17, 31, 72, 183
expenditure *see* funding

Farmbox Meats 10, 12, 87, 98
fat burner, DNP as 108, 109, 110
fatalities *see* deaths
fatty acid profile analysis 185
FDA (Food and Drug Administration) 90, 91–2, 214
Federal Food, Drug and Cosmetic Act (1938) 91, 109
Fibonacci spiral 23
financial (banking) crisis (2008) 139–44, *see also* costs; funding; prices
Findus 26, 147–8
fipronil in Dutch eggs 62–6
fish 78–82
   restrictions and pressures on fishing 130
   supply chains 130
$5 \times 5 \times 5$ intelligence recording systems 194

FlexiFoods Ltd 14, 15
fluoridation of water 176–7
Food and Agriculture Organization (UN) 82
Food and Drug Administration 90, 91–2, 214
Food Hygiene (Wales) Regulations (2006) 87
Food Hygiene Rating System (FHRS) 152, 163, 179, 208
Food Industry Intelligence Network (FIIN) 30
Food Law Enforcement Practitioners (FLEP) 120, 198
Food safety *see* safety *and entries below*
Food Safety Act (1860) 99, 164, 198
Food Safety Act (1990) 13, 52, 64, 89, 95, 96, 121, 124, 166, 199
Food Safety Agency, European (EFSA) 63, 66, 90, 94, 168, 208
Food Safety and Hygiene (England) Regulations (2013) 87, 88
Food Safety Modernization Act (FSMA) 91, 92, 93
Food Safety Standard (BRC) 37, 38
Food Standards Act (1999) 164
Food Standards Agencies (FSAs)
    European (EFSA) 4, 90, 94, 164–5, 168, 169, 206
    Ireland (FSAI) 5, 6, 7, 16, 130, 149
    Scotland (Food Standards Scotland) 30, 31
    seeking help from 135
    UK 6, 7–8, 21, 27, 29, 30, 31, 35, 39, 40, 141, 147, 148, 172, 174, 199, 201, 206
        expenditure reductions 172, 206
        Findus and 147
        industry seeking help from 135
        National Food Crime Unit (NFCU) of 28, 38, 40, 42, 133, 136
        prosecutions and 9, 10, 12
        roles and expectations 164, 165, 177, 207
Food Standards Scotland (FSS; Food Standards Agency Scotland) 30, 31
forensic accounting 213–14
forensic science and scientists 199, 201
Fourier transform infrared spectroscopy 186, 187
France (the French)
    horsemeat 171
    wine scandal 75–8
fraud
    costs *see* costs
    definition 54, 120–1
    five elements demonstrating 3, 54
freedom of movement of food across borders, EU and 207, 216
Freeza Meats Ltd 5
French, the *see* France

frontier ports 154
funding (and investment/budgets/resources relating to enforcement) 200-1
    cutting/falling/withdrawal/lack of 199, 201, 202, 205, 206
    local authorities 113, 142, 145, 146, 158, 167, 168, 171-2, 179, 187, 188, 199, 206, 211
    specialist enforcement teams 133
fungal toxins (mycotoxins) 153
future (enforcement and regulation)
    ensure prosecution success 98-9
    fears for 117
    predicting 99, 189, 190-7, 213-14

game theory 124-5
Garda National Bureau of Criminal Investigations 17
General Food Regulations (2004) 11, 87, 88
genetic modification 90
geographic protection status 71, 166
geographical origin, testing for 35, 184
Germany 154, 155-8, 200, 207
    asparagus 66-7, 173, 184, 194
    as gatekeeper of UE food safety 154, 155, 159
    laboratories and analysts 34, 35
        qualifications 199
    UK sharing and working together with 35, 36
    vegetables contaminated with *E. coli* 174
gin 48, 123
Giuliani (New York mayor) and zero tolerance to crime 190-1
global dimensions *see* worldwide dimensions
Global Food Safety Initiative (GFSI) 214
Global Standards (BRC) 37, 38
goat meat labelled as lamb 12-13
Goodwin, Fred (Fred 'the shred') 140
government (generally or non-UK at national level) and politicians 133-4, 202-3
    EU and safety enforcement and the EU 164-5
    key stakeholder role in food safety 176-7
    media reporting on non-compliance and reactions by 173-5
    Spain and olive oil scandal 73-4
government (local) *see* local authorities/government
government (UK national) 86-7, 133-4
    business and 134, 141, 142-3, 179, 216
    in Elliott Report 23, 29, 43, 200-1
        crisis management 41, 42

intelligence gathering 29–30, 31
laboratory services 32, 33–4
support 39
learning from banking crisis 142–3
media and 173–5
Greek olive oil 71, 72
growing and harvesting stage, vulnerability at 129
Gumpert, David E. 116

Hampton review 167, 169
harvesting stage, vulnerability at 129
Hassall, Arthur Hill 46, 47–8, 122, 123, 163
Head, Steve 124–5
health and safety legislation 178
Hill, Dr Alfred 50
histamine 79, 80
historical perspectives 44–59, 121, 121–3, 163
adulteration 45–59, 62, 121–3, 163–4
home authority principle of enforcement 167–8
horse DNA testing 6–7, 9, 16, 17
Horseferry Road Magistrates Court 95
horsemeat 9–18
legal trade 5
substitution of beef with (and investigation and prosecution) in UK 9–18, 86, 124, 143, 147–9, 171, 211, 213
cases 9–16

companies involved 26, 147–9, 150
Iceland's CEO on 211
key stakeholders 177
public impact 25
Hungary 200
hydrogen peroxide 78, 79
hygiene (regulations) 87, 88–9, 142, 169
hygiene service/enforcement 165, 178, 201, 203

Iceland (supermarket) 137, 211, 212
IdentiGen (Ireland) 6
*i*Knife 185
illicit drugs, mimics 111–12
immuno-gel 185
in the public interest 55–6
independent national services need 134, 179–80, 202–3
India, chilli powder 82, 94, *see also* BRICS economies
industrial oil contaminating olive oil 72–3
industry *see* businesses
infant (baby) milk, melamine 21, 22, 68–71, 117, 189, 199–200
information *see* communication; documentation and record-keeping; emerging data; intelligence; sharing
infrared spectroscopy 186, 187

innovation/new systems
  183–204
    enforcement 180, 183–204,
      208
    need 183–4
    technology and techniques 34, 35, 184–9, 200
insecticides *see* pesticides
Institute for Global Food
  Security (Belfast) 22, 31
insurance 139, 145, 146
integrity (food)
  Elliot Review *see*
    Elliot Review
  untargeted or fingerprint
    approaches 186
Intellectual Property Office,
  EU 166
intelligence (information)
  29–31, 189–91, 197–8,
  see also communication
    challenges to using 197–8
    continuum 193
    establishing systems of
      134
    5×5×5 recording
      systems 194
    gathering 29–31
    how to use 189–91
    sharing *see* sharing
intelligent knife (*i*Knife) 185
International Meat Traders
  Association (IMTA) 38
international scale *see*
  world-wide dimensions
Internet (online and
  world-wide web) 103–4
    data sharing research
      194–5
    diet pills 108

Internet of Things 190
TSB banking fraud
  150–1
Interpol 31, 194
investment
  banking crisis and 139
  in enforcement *see*
    funding
Ireland 2, 16
  Food Standards Agency
    (FSAI) 5, 6, 7, 16,
    130, 149
  meat 5–6, 7, 14, 15, 16
irradiation 89–90
isoelectric focussing
  electrophoresis 185
isotope analysis
  isotope ratios 35, 67, 184
  in SNIF-NMR® 184
Italy
  fish bleaching 79
  as gatekeeper of UE food
    safety 154, 155, 156
  olive oil 71, 72, 75

Jack, Lisa 212–13
James, Christopher (lawyer) 11
John (King of England) 45,
  51, 56, 57

Kantar Worldpanel 25
King John 45, 51, 56, 57
Kratom 110–11
Kurniawan, Rudy 76, 77

labelling, false and misleading
  avoiding 88
  DNP 110
  fish 80, 81
  horsemeat 12, 17

Subject Index

laboratory services (analysts) 31–7, 135–6, 200–2
   competition *see* competition
   historical perspectives 46, 47, 49–50
   private sector 32, 33, 200–1, 202, 203
   public sector 32–4, 201, 202
   referee analyst 8, 33, 202
   seeking help from 135–6
lamb, falsely/misleadingly-labelled 12
Lataste, Vincent 76
law *see* legislation
Lawrence, Felicity 35–6
lead 45, 48, 57, 123
leadership 40–1, 180, 208
   strategic 208
learning (and its improvement) 208, *see also* training
   from banking crisis 139–44
legal highs 111–12
legislation (law) and regulations 86–119, 120–5, 162–6, *see also* enforcement; self-regulation
   banking crisis and lessons to be learned by regulators 138–43
   businesses and 88, 146, 176
   changing 133, 162–6, 202
   complex 163–5
   future *see* future
   General Food Regulations 2004, regulation 4 11, 87
   key stakeholder role of legislators 175–6
   supplements 105–10
   world-wide harmony and consistency in 89–90, 93, 94, 177
      lack 89–90, 165–6, 197
Letheby, Dr Henry 52, 62, 93
Leucogee (alum) 45, 47, 122
LGC (Government Chemist) 36, 37
Lidl 7
Liverseedge, J.F. 50
local authorities/government (and role in enforcement) 35, 36–7, 93, 96, 113, 158, 165, 167–8, 172, 177, 179, 201, 203, 211
   funding/budgets/resources 113, 142, 158, 167, 168, 172, 179, 199, 206, 211
   historical perspectives 49
   officer training 99
Lutjanus purpureus 80

McAdams Foods 5, 14
McCreath, Alistair (judge) 11, 13
Macroscope 195, 196
Magistrates Court 2, 22, 55, 87, 95, 100, 143
Mansion house speech by Gordon Brown 140
market (on-the market) surveillance and monitoring 5, 98, 106, 113, 155
Marks and Spencer 211–12
Marret, Francois-Marie 75–6

Mastership in Chemical Analysis (MChemA) 199
Mayne, Sir Richard 189
meat, *see also* abattoirs and slaughterhouses; beef; chicken; goat meat; horse; lamb; pork
    classification 217–18
    E. coli O157-contaminated 96–7
    hygiene service/enforcement 165, 178, 201, 203
    minced 2, 3–4
    supply chains 38
meat trade/industry
    how to commit fraud 3–5
    International Meat Traders Association (IMTA) 38
    investigation 5–7
mechanically separated or recovered meat 4
media 136, 173–7
    breaking news 77–8, 194–5
Medicines and Healthcare Products Regulatory Agency (MHRA) 105, 111, *see also* drugs and medicines
medieval period (Middle Ages) 45–6
melamine
    Chinese baby milk 21, 22, 68–71, 117, 189, 199–200
    pet food 214
Mens Rea 22, 98, 121
merchant guilds 45–6
Middle Ages 45–6

mid-infrared spectroscopy 186
milk
    baby, melamine 21, 22, 68–71, 117, 189, 199–200
    watering down 49–50, 58
mimics
    of illegal drugs 111–12
    of wine, legal 77–8
minced meat 2, 3–4
mislabelling *see* labelling
Mission Impossible (author's paper) 104
Misuse of Drugs Act (1971) 112
mitigation 54–6
Mitragynine speciosa 110–11
monitoring 96, 97, 116, 171, *see also* surveillance
    of emerging trends 195
    EU countries 155, 158
    local authorities 211
    risk-based 6, 173
Moreau, Luc 197
mortalities *see* deaths
Moss and Boddy (case) 9–12, 87, 100
mouse contamination 95
Multi-Annual National Control Plans (MANCP) 168–9
Muro (Dr) and Spanish olive oil scandal 73, 74
mycotoxins 153

NAS (Italian Food Police) 79
national bodies/services, independent 134, 179–80, 202–3
National Crime Agency (NCA) 40, 126

National Food Crime Unit
  (NFCU) 28, 38, 40, 42,
  133, 136
National Intelligence Model
  (NIM) 189, 191
Naughton, Declan 127, 151
near-infrared spectroscopy 186
Nepusz, Tamas 151
Netherlands (the Dutch) 114,
  116, 207
    eggs containing fipronil
      62–6
    water addition to
      chicken 188–9
network analysis tool 127, 151
new systems *see* innovation
New York, zero tolerance to
  crime 190–1
news *see* media
NExT++ 195
Nielsen, Ulrich 14, 16, 100
nitrates 79, 80
non-compliance 56–8, 114,
  127, 163, 170, 174, 198
Northern Ireland,
  horsemeat 15
nutrition, sports *see* supplement

Oceana 80
Official Control Laboratories
  (OCL) 32
oil (industrial) contaminating
  olive oil 72–3
olive oil 48, 71–5, 122, 123
on-the-market monitoring
  (market surveillance) 5, 98,
  106, 113, 155
on-the-street monitoring
  (street surveillance) 6,
  158, 215

one-size-fits-all (Procrustean
  bed) approach 114, 116,
  198–200, 201, 203
online *see* Internet
Operation Opson 31, 40
opportunistic fraudsters
  126–7, 128
Opson 31, 40
oregano 41
organised fraudsters 72, 126,
  127, 183
organophosphates 74
Ostler-Beech, Alex 14, 100

packaging stage, vulnerability
  at 129
pargo 80
partnership 4, 180
    public sector 33–4
Paterson, Owen (Environment
  Secretary) 147, 148
Patterson and Raw-Rees
  (case) 12–14
PCR (polymerase chain re-
  action) test 8, 186
peppers 213–14
    cayenne 48, 123
Perrier bottled water 149
pesticides (incl. insecticides)
    Dutch eggs 62–6
    peppers 214
    tomatoes 74
pet food, melamine 214
Petroczi, Andrea 127, 151
Piccinno, General Cosimo
  79
pickles 48, 123
pigs *see* pork
Pliny (AD70) 45, 122
point of sale *see* retailers

police 189–91, 194, 197, 201
    olive oil fraud 71
    UK 98, 99, 187, 189–90, 201
Polish beef sold as British 14, 28
politicians *see* government
Polnay, Jonathan (prosecuting barrister) 15
polymerase chain reaction (PCR) test 8, 186
Ponsot, Laurent, and Domaine Ponsot 76–7
pork (from pigs/porcines)
    DNA testing 6, 7
    substitution of 2
port(s) 113, 114, 154, 177
    frontier 154
porter 48, 123
Portsmouth University 212–13
poultry farms and Dutch egg contamination 64
prediction of crime/fraud in future 99, 189, 190–7, 213–14
pressuring of fraudsters 135
Preuβ, Dr Axel 67
prevention *see* defence
prices and value, commodity/food 130, *see also* costs; funding
    Brexit and 216
    high/premium 71, 75, 80, 132, 196, 213, 217
primal cuts of meat 3
Primary Authority Partnership 167–8
prison *see* custodial sentences
private sector laboratories 32, 33, 200–1, 202, 203
Proceeds of Crime Act 133

processing stage, vulnerability at 127–30
Procrustean bed approach 114, 116, 198–200, 201, 203
product of designated origin (PDO) 78, 130, 132, 213
    olive oil 71
professional qualifications 36
profitability of fraud 132
prosecutions *see* court proceedings
protected geographical indication (PGI) status 71, 166
protection *see* defence
provenance 133, 195–7
public (the public), *see also* consumers
    adulteration and 50–1, 94
    horsemeat and 5, 166
    in the interests of 55–6
    involvement 208
        crime reporting 191
    media reporting on non-compliance and reactions by 173–5
Public Health Security and Bioterrorism Preparedness and Response Act (2002) 91–2
public sector analysts/laboratories 32–4, 199, 200, 201, 202

QualiBordeaux 76
qualifications *see* training and education

Raman spectroscopy 186–7
Rangelands 15, 17

Rapid Alert System for Food and Feed (RASFF) 52, 63, 65, 70, 93, 106, 107–8, 109, 111, 113, 114, 122, 123, 127, 158, 167, 170, 172, 206–7, 213
   author and 151
   top 3 gatekeepers and 155–6
Raw-Rees and Patterson (case) 12–14
RBS (Royal Bank of Scotland) 140
record-keeping *see* documentation and record-keeping
red snapper 80
referee analyst 8, 33, 202
registration of businesses 91, 92, 180
regulations *see* enforcement; legislation and regulations
Replica Wines 77–8
resilience 32, 208
   UK 132
resources *see* funding
retailers (point of sale)
   high street, and the Internet 103, 104
   vulnerability 129
Rhomboplites aurorubens 80
rice, melamine 69
Richards, Alan 55
risk, and its perception 132
risk-based systems
   enforcement 141, 168–70, 198
   monitoring 6, 173
rodent contamination 95
Romans 45, 57, 122

Royal Bank of Scotland (RBS) 140
Russian roulette (author's paper) 104, 109

safety 145–61, 177–9
   dealing with unsafe food 88
   enforcement *see* enforcement
   gatekeepers 151–8
   key stakeholders 174, 175–7
   responsibility for failures 177–9
Sale of Food Act (1582) 45
Sale of Food and Drugs Act (1875) 36
salmon 80
Salmonella 61
Schmidt, Professor (University of Munich) 67, 184
school meals 97, 124, 135, 197, 211
Scotland
   Food Standards Agency/ Food Standards Scotland (FSS) 30, 31
   laboratory testing 201
self-accreditation 180
self-regulation 143, 178
sentences *see* court proceedings; custodial sentences
sharing of data/information (collective intelligence) 30, 31, 35, 36, 133, 134, 189, 194, 195, 197, 208
   research 194–7
   sensitivity issues 197
shortages of product 213–14

Shotton, Stuart 72
Sideras, Andronicas 14, 15, 16, 100
sildenafil 111–12
Silvercrest 6, 7, 15, 16, 17
Singapore University 194–5
Skelt wine 184
slaughterhouses and abattoirs 4–5, 8, 9–10, 12, 165, 171
Smollet, Tobias 50
snapper 80
SNIF-NMR® 184
Snow, Dr John 47
SOCIAM research 195
Southampton (Univ.)
    ADHD and 'Southampton Six' (colours) 94
    school meal supply chain mapping 197
    SOCIAM research team 195
Southwark Crown Court cases of horsemeat fraud 9–16
Spain
    horsemeat 17
    olive oil 71, 72, 72–5
Spanghero 171
Spanish Fly 215
specialist enforcement teams 121, 133, 183
spectroscopy 186–7
Spencer, Timothy 95
spice merchants 46
Spink, John 121, 214
spirit fraud, costs 166
sportpersons and athletes 105–11
stakeholders in food safety 174, 175–7
Standard for Agents and Brokers (BRC) 37

Steed, Sara 125
stout 48, 123
strategic leadership 208
street surveillance (on-the-street monitoring) 6, 158, 215
strict liability 22, 89, 121
substitution
    of beef *see* beef
    by chicken 60–2
    definition 121
    by horsemeat *see* horsemeat
    of pork 2
Sudan dyes 21, 82, 94
sulfuric acid 48, 57, 123
supplements (sports products inc. nutrition) 104, 105–11
    definition 105–6
    as future soft target 213
supply 41, 140
    chains/webs 130, 170–1, 210
        complex/convoluted 41, 42, 170–1, 212
        fish 130, 135
        meat 38, 42
        shorter 4, 134–5, 212
        world-wide 170–1
    relationship with suppliers 130
    shortages in 213–14
surveillance 33, 42, 107, 113, 133
    market (on-the-market monitoring) 5, 98, 106, 113, 155
    street (on-the-street monitoring) 6, 158, 215
synthetic food 51–2, 93–4

Takatsrial, Zoltan 185
Tavola 171
tea 46, 48, 51, 57, 111, 123
technology and techniques,
   new/innovative 34, 35,
   184–9, 200
tendering 203
      competitive 34, 201, 205
Tesco 6, 7, 9, 16, 26, 39, 148–9
Thai chicken 188
thinking like a
   criminal 120, 208
tomatoes,
   organophosphates 74
toxic oil syndrome 72–5
training and education (incl.
   qualifications) 41, *see also*
   accreditation; learning
      auditors 38, 39
      investment downturn
         201, 205
      local government
         officers 99
      public analysts 34, 199
transgression by nations 127,
   128, 152, 180
transition economies 70, 104,
   170–1
trim (meat) 3–4
Troop, Pat 174
Truss, Elisabeth 86
TSB bank fraud 150–1
Tsinghua University 194
tuna 79–80
Turkish nuts 95

United Kingdom (UK/Britain),
   see also England; Northern
   Ireland; Scotland; Wales
      borders 158–9, 170, 216

   chicken, water addition
      187–8
   Elliot Review *see*
      Elliot Review
   as gatekeeper of UE food
      safety 154, 155–6, 157
   Germany and, sharing in-
      formation and working
      together 35, 36
   government *see* govern-
      ment (UK)
   historical perspectives *see*
      historical perspectives
   horsemeat *see* horsemeat
   law and regulations 121–5,
      169–70
      US law compared 90
   LGC (Government
      Chemist) 36, 37
   perception of 'burglar
      alarm' on 132–3
   police 98, 99, 187, 189–90,
      201
   Polish beef sold as British
      beef 14, 28
   reduced spending on
      enforcement 172
   signs of progress 136–7
   sources of imported
      food 154
   supplements market 107
United Nations Food and
   Agriculture Organization 82
United States (USA)
   fish 80, 82
   Food and Drug Adminis-
      tration 90, 91–2, 214
   law 91–2
      compared with
         UK 90

unsafe food, dealing with 88
USA *see* United States

value *see* prices and value
Viagra 111–12
'Victorian Bakers' (documentary) 50, 124
vinegar 48, 57, 123
Von Neumann, John 124
vulnerability (food) 127–30
    assessment 37
    fish 82

Wakley, Thomas 46
Wales
    Food Hygiene (Wales) Regulations 2006, regulation 17(1) 87
    horsemeat testing 143
    public analysts 201
Walker, Caroline 51, 52
Walker, Malcolm (Sir) 211, 212
Walker, Michael 8–9, 36, 185
water
    addition to chicken 187–9
    addition to milk 49–50, 58
    bottled, benzene in 149
    chlorinated, washing chicken 90
    fluoridation 176–7
wheat
    historical perspectives 45, 122
    melamine 69

Which (consumer magazine) 25, 26
whistleblowing 28, 29–30, 134, 208
WHO (World Health Organization) 62, 164, 165
wholesaler vulnerability 129
Wilson, Bea 51, 52, 117
wine
    costs of fraud 166
    French 75–8
    German (Skelt) 184
    Roman times 45, 57, 122
Winston, Robert 187
Wood, Sir Charles 46
Worcester sauce 82
Workman, Tim (Stipendiary Magistrate) 54, 95
World Health Organization (WHO) 62, 164, 165
world-wide dimensions 60–85, *see also* Global Food Safety Initiative; Global Standards
    customer buying from other side of the World 212
    harmony in food law *see* law
    supply chains/webs 170–1
    transgression by nations 127, 128, 152, 180
world-wide web *see* Internet

X-ray diffraction techniques 186–7

zero tolerance 28–9
    New York 190–1